# THE YEAR OF THE PYRES

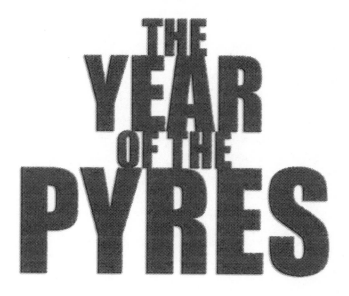

# The 2001 Foot-and-Mouth Epidemic

# Judith Cook

MAINSTREAM
PUBLISHING

EDINBURGH AND LONDON

First published in Great Britain in 2001 by
MAINSTREAM PUBLISHING COMPANY (EDINBURGH) LTD
7 Albany Street
Edinburgh EH1 3UG

ISBN 1 84018 550 3

A catalogue record for this book is available from the British Library

Typeset in Bembo
Printed and bound in Great Britain by Mackays of Chatham

# Acknowledgements

A good many people have assisted me by providing information and research help for this book, in particular Jane Mayes who has checked out information for me and undertaken interviews in the north of England, which sheer distance precluded my doing in the time available. My thanks to Richard Mawdsley for continual updates on the situation in Cumbria from a farmer's viewpoint and for permission to use excerpts from his journal, and to Peter Poole from Hereford for keeping me informed on the situation there. Also to my many e-mail correspondents from all over the country, but particularly in the West Country. I very much appreciated the interview given to me by Anthony Gibson, Director of the South-west region of the NFU, whose own weekly column in the *Western Morning News*, and regular television appearances, have provided sanity in the midst of chaos. Also my thanks to James Bell for his information on pollutants and for permission to use his paper on pollution from pyre smoke as an endpaper. Also to all those who have patiently answered my questions and given me information.

I feel there should be a special mention for the regional press which, unlike the mass media, has kept its eye on the ball and never let the story slip. My own local daily paper, the *Western Morning News*, has been exemplary in its outstanding coverage, and is also responsible for the

Heart of Devon campaign, which has now been taken up all over the country. My special thanks to the librarians on the *Western Mail* and the *Yorkshire Post*.

Finally, I much appreciated the letters from Jon Snow and Ian Hislop encouraging me to go ahead with this book.

# Contents

# Foreword

It was a beautiful spring day just before the Easter weekend and I had just driven us through Tewkesbury, with its picture postcard timbered houses and narrow streets. We drove a short way out of the town and on to a main road. On the horizon ahead of us lay the Malvern Hills, blue and misty. The hedges were full of blossom and along the roadside there were drifts of daffodils: an idyllic scene, the countryside at its best. A mile or two out we came to a 'Road Closed' notice. To our left was a flooded field. Nothing surprising in that, the flood plains around Tewkesbury, Worcester and Gloucester regularly flood when the River Severn is in spate and the situation had been worse than usual during the previous months. But the road was clear. I stopped the car.

Then a policeman appeared. I asked why I couldn't continue on my way since there was nothing on the road. It was, he explained delicately, 'a public health matter'. I asked what he meant. It was because of foot-and-mouth. There had been a cull of animals on a nearby farm, he said, several hundred of them. Unfortunately, while they were lying waiting for disposal, the river had flooded and covered them up. The authorities were trying to work out what to do next. They wanted to avoid a pollution problem but it was very difficult with all that water. And now the carcasses were starting to rot.

So we turned round and went another way. But when I got back to

Cheltenham, where we were staying, I watched both the BBC and ITN News, assuming that such a major event was bound to feature, since both channels were running stories on the epidemic. There was nothing. The next morning, before driving back to the West Country, I bought several national newspapers. Still nothing. It was as if I had dreamed it.

It was after we passed Exeter that the smell – soon to become familiar – hit us, from the pyres on each side of the road. It was the stench of burning animals. I had been following the story since the very beginning but it struck me forcibly that, setting aside the ignorance, conflicting advice and downright misinformation peddled by the authorities, people in one part of the country simply did not know what was going on in another. So I wrote a letter to *The Guardian* to that effect.

The next day my computer groaned under the flood of e-mails from all over the country. My surmise had been only too true. For several weeks I found myself running an unofficial information exchange, until pressure of other work forced me to stop. At least by that time I had learned of the various websites set up for that purpose.

As Jon Snow of *Channel 4 News*, wrote in *The Guardian* on 25 April 2001:

> Is the great foot-and-mouth disaster rapidly becoming the first e-mail democratising story of our time? Against the background of MAFF's [the Ministry of Agriculture, Fisheries and Food] self-interested manipulation of figures . . . the gulf in information from 'the field' has been filled in by people directly affected by the consequences of the disease . . . These are not the propagandist circulars from the Soil Association or other pressure groups, but person to person e-mails, coming apparently at random, from as far afield as Dumfriesshire and Cornwall. They have inevitably played an increasing role in shaping our approach to covering this catastrophe.

This book, therefore, sets out to tell the story of the worst epidemic of foot-and-mouth disease the world has ever seen. The slaughter, at the last count, of nearly four million animals and rising (many of them quite healthy); the virtual closure of the countryside for months; continuing fears of long-term pollution of the environment from pyres and burial pits; a cost to the Treasury of billions; the postponement of a general election and, not least, the traumatic experiences of some of those who have been so severely affected.

At the time of writing, although we are constantly being told that the epidemic is now 'tailing off', and has been controlled, it is not over yet. The worst case, the nightmare scenario, sees it continuing well into next year and the possibility that the virus will never now be eradicated from our farming stock.

\* \* \*

It is a well-known truism that those who do not learn from history are destined to repeat it. It is absolutely essential that we learn from this catastrophe, but the omens are not promising.

On 13 June 2001 the BBC's *Today* programme ran an interview with Jose Bove, the French farmer who become a national hero after being sent to prison for assisting to trash a half-built branch of McDonald's in the small town of Millau. He was over in England helping to publicise the book he has written with François Dufour, *The World is Not for Sale: Farmers Against Junk Food*. Was it our farming methods which had contributed to our epidemics first of BSE, and now of foot-and-mouth disease, asked the interviewer?

In his careful English, Monsieur Bove replied that he thought it probably was. The 'liberalising' of regulations had led to the manufacture of the cattle feed which had caused BSE in cows; and the way animals in England were transported on long journeys between livestock market and farms, and farms and abattoirs, had ensured that the foot-and-mouth virus spread out of control. Asked to comment on this, Ben Gill, CBE, President of the National Farmers' Union and a major player in this story, answered with one word:

'Rubbish!'

# Chapter 1

# Pigswill

The Year of the Pyres began very quietly. On 19 February 2001 a young vet, routinely checking animals at Cheale Meats, an Essex abattoir, discovered signs of disease. He was too young to have been involved in the last serious outbreak in 1967 but the symptoms fitted those for foot-and-mouth disease. Blood tests confirmed his diagnosis the next day.

The source was reportedly tracked back to a pig-fattening outfit at Heddon-on-the-Wall in Northumberland close, as its name suggests, to Hadrian's Wall. The unit (described by the local MP as 'a hill-billy outfit') was run by Bobby Waugh and his brother, neither of whom lived on site, and it transpired that during the previous winter there had been a number of complaints about the state of the place. Witnesses spoke of seeing limping animals, dead animals lying around for days, and untreated pigswill. Complaints had been made to Newcastle Trading Standards officers, environmental health inspectors, the local branch of MAFF and the RSPCA who, after viewing the place from outside (as they were refused entry), made two formal complaints.

Belatedly, local council health officials took a look (their report is still unpublished at the time of writing), as did representatives from Newcastle MAFF. The latter did not much like what they saw, but felt it was all right for the brothers to carry on, at least for the present. Attention immediately focused on the pigswill fed to the pigs by the

Waughs. Only a minority of pig farmers still used this method and strict guidelines had been laid down for its use, the most important of which was that swill must always be boiled before being fed to pigs. The Waughs were licensed to collect waste food from a variety of sources including schools, hospitals, and restaurants. A suggestion that the source of contaminated food at Heddon-on-the-Wall was specifically waste from Chinese restaurants was furiously refuted by Chinese restaurateurs and led to public apologies. Whatever the source, however, allegations were made to authorities in Newcastle that the swill had not been processed properly and that pieces of half-cooked chicken had been seen in the swill.

Later a Lincoln-based MAFF vet, Charles Ross, claimed that MAFF and the government had been aware of the 'significant risks' of feeding swill to pigs for some time before the outbreak of the epidemic. In a letter written in April and quoted in the *Western Morning News*, he stated that there had not been a single occasion during his inspections when he did not come across serious breaches of the regulations and that welfare problems existed on at least half the farms he visited. It was MAFF policy, he wrote, to 'try and encourage offenders to improve rather than prosecute'. It was, he claimed, common knowledge that pig farming practices were a 'time bomb waiting to go off'. However, it was not until 24 April 2001, over two months after the start of the epidemic, that the government finally banned the feeding of catering waste to pigs.

But whether or not the Waughs had properly treated their swill, the virus did not originate at Heddon-on-the-Wall. The specific strain of the virus, Pan Asiatic O, had first surfaced ten years ago in the Far East and had been spreading towards the West ever since, a fact of which the authorities must have been fully aware. There is also a possibility, which will be examined later, that the virus was already present in sheep before the Heddon outbreak. It is not necessary to go into detail about the theories that the disaster was due to sabotage by a crazed eco-warrior, or a stolen phial from the government's biological warfare research station at Porton Down, though both have been posited. Experts believe it is far more likely that it entered the food chain legally in contaminated meat.

As late as the end of 2000, Britain was still importing meat from countries where foot-and-mouth disease is endemic such as Argentina, Botswana and Namibia. It came as something of a shock to learn that this was the case. The common perception is that we are virtually self-supporting where meat is concerned, apart from the importation of New Zealand lamb, which is a special case. But many commentators have made the point that this is one of the results of global trading agreements and that such is the power of the supermarkets, which buy in huge quantities and effectively dictate prices, that the 2001 foot-and-mouth epidemic was an accident waiting to happen. As this imported meat was cheaper than that produced at home, it often ended up in school meals and army canteens and could have got into pigswill through this route. On the other hand, the virus might even have come in to the country in a pie or a sandwich bought abroad to eat on the way home or from a not-too-fussy local shop.

\* \* \*

At this point it may be helpful to introduce the principal organisations and people who were to play a major role in the unfolding story. First under the spotlight is MAFF, the old Ministry of Agriculture, Fisheries and Food (replaced by DEFRA – Department for Environment, Food and Rural Affairs – in June 2001). Even before the disaster of BSE and the outbreak of swine fever in 2000, MAFF had long been criticised for what appeared to be its lack of interest in, and care for, food consumers. It was widely seen as being run almost solely in the interests of farmers in general and the big agribusiness barley barons of East Anglia in particular.

If the office of Minister for Northern Ireland is considered a poisoned chalice, that of Minister for Agriculture was (and possibly still is) the equivalent of the ministerial wooden spoon. Civil servants with ambitions steered well clear of MAFF, jokingly describing its headquarters as 'the graveyard'. The previous Conservative administration had been quite content with the situation, brushing aside all suggestions of reorganisation or the setting up of a new department altogether. Indeed, it had seen MAFF as an area in which economies could be made with the result that, during the years of the Conservative government, over 200 vets employed by the Ministry were made redundant, leaving in post only half the number employed 20 years ago. This was later to prove just one of a long series of costly errors.

Even as late as the general election campaign of June 2001 Tim Yeo, who had been Shadow Agriculture Minister, was refusing to be drawn on the future of MAFF, when both the Labour Party and Liberal Democrats were campaigning for a new department altogether. Possibly this had something to do with the fact that fields in Dumfries & Galloway and in Devon, two of the worst affected areas, sprouted a rash of huge 'Vote Conservative' posters in areas closed to walkers and the general public, where even lay-bys were cordoned off and there were disinfectant mats across the roads. He would have had no wish to rock the boat at such a sensitive time.

The unfortunate incumbent at the Ministry when the foot-and-mouth outbreak started was Nick Brown, ironically enough, Labour MP for Newcastle upon Tyne and Wallsend. Brown is described in *Parliamentary Profiles* as 'a Kent-born Geordie; dark, stocky, full-faced, pointed nose and flat, parted brown hair . . . a thick-set fellow who, at first sight, resembles the stud-collared bulldog called "Spike" in the Tom and Jerry cartoons'. He first arrived in the Commons in 1983, became a spokesman on a variety of issues and served on a number of Select Committees, none of which had anything to do with agriculture. He had a reputation for toughness and in 1997 was made Chief Whip.

In December the same year, however, he replaced Jack Cunningham as Minister for Agriculture. According to the Westminster rumour mill he was shuffled into the graveyard slot of Minister for Agriculture because, although he was 'the backroom fixer' who helped organise the Brown and Blair campaigns for membership of Neil Kinnock's Shadow Cabinet in the late 1980s, he then blotted his copybook by going on to assist Paul Routledge with his biography of Gordon Brown; thus helping to publicise the continuing Blair–Brown tension. He has also been described as a staunch believer that the role of individuals should be subservient to the good of the party: which might explain why, however much pressure he might have been under, he never once wavered from the agreed government line.

Then there was the National Farmers' Union, the NFU, under its combative President, Ben Gill, again often viewed from outside as the voice of 'Big Farming'. Only 30 per cent of farmers actually belonged to the NFU at the start of the outbreak and at no time, while the epidemic was at its height, did it call a council meeting to discuss or ratify the policies of its leaders. This was to prove extremely divisive, although, to be fair, some of its regional officers were far more sympathetic and open to discussion, particularly those in the south-west which has suffered so much.

Last, but not least, was the Prime Minister, Tony Blair, assisted by his advisers. Following the election victory of 1997, the new Labour administration had geared all their efforts towards winning a full second term in office, something no previous Labour government had ever achieved. The countdown to the second election was almost militaristic in its precision and it was an open secret that the Blair administration favoured 3 May, which was also the day on which county council elections were to be held in England and Wales. Even in their wildest nightmares the Blair administration could never have imagined what was going to hit them.

* * *

On 23 February 2001, foot-and-mouth disease was officially confirmed at Heddon-on-the-Wall. Worse, there was every reason to believe that it had been present in the Waugh pigs for at least three weeks. This, one imagines, should have been the time when MAFF dusted off the 1968 Northumberland Report into the last major outbreak of foot-and-mouth. But in that outbreak, serious as it was, the disease was limited to counties in northern England and the Midlands, with Cheshire coming off worst. Some 430,000 animals were slaughtered and the resultant carcasses dealt with. The report states that experience showed that burial was preferable to burning 'because of the relative speed of the operation'. The delay caused by preparing pyres added to the risk of spreading the disease, not least because the virus might be carried elsewhere on thermal air currents. The whole tenor of the report was that speed was of the essence in order to combat the disease: speed of diagnosis and speed of action after diagnosis was confirmed. It concluded that 'our proposals for pre-outbreak planning and for an organisation capable of rapid expansions to deal with epidemics should reduce delay to the minimum in the future'.

It is now clear, however, that little or no notice was taken of the burden of the report or its conclusions as when the disease resurfaced in 2001 no contingency plan existed. As a former vet, Ken Tyrrell, who diagnosed the first and last case of the disease 30 years ago told the *Western Morning News* on 26 April, 'At the end of the day we recorded our findings for the benefit of future generations. What the hell for? No one has paid the slightest attention.'

Once the presence of the disease had been officially confirmed, MAFF announced an immediate ban on livestock movements. But the three-week delay before it became known that the disease was in the country, followed by the three-day gap between the discovery of the first case in the Essex abattoir and its confirmation at Heddon-on-the-Wall, meant that such a ban had come far, far too late. And, as was to become all too obvious, MAFF and the government then grossly underestimated the number of animal movements during that crucial period.

When it was later discovered that the rapid spread of the disease had been caused by sheep, first estimates put the number of sheep movements over the three to four weeks prior to 23 February at about 2,000. Later, when MAFF had given up on trying to trace all possible sheep movements, it was put at 1.35 million. If the senior officials in MAFF had been aware of this most basic piece of information it might, it just might, have been possible to check the disease then – if sufficient expertise, time, effort and, above all, speed had been applied. As it was, by the end of the first week of the epidemic it was already too late.

# Chapter 2

# Sheep, Sheep Come Home

Foot-and-mouth disease first appeared in this country in 1839 and, although it caused concern, it was largely ignored for the next 30 years, as it was common, extremely mild in relation to other prevalent diseases, and provoked few efforts at control. It was an accepted, indeed expected, occupational hazard, as is explained by Abigail Woods of the Wellcome Unit of the History of Medicine, Manchester University, in her paper, 'Evaluating the Current Control Policy from a Historical Perspective'. Attempts at control by the state came much later as initially attention was primarily focused on preventing the importation and spread of diseases such as cattle plague that were considered to be much more serious. Most vets, farmers and MPs saw no particular need to try and control foot-and-mouth, since the disease was not particularly severe, most animals recovered naturally and losses due to a legislative 'cure' through slaughter would outweigh those inflicted by the disease itself. They also doubted if control was possible, as they believed it might be spread by human and wildlife movements, which could not be controlled as easily as infected livestock. As far as is known, it is not transmissible to humans.

The virus is shed from the respiratory tract in the form of a small particle aerosol less than five microns in diameter and then suspended in the air; therefore it can be transmitted passively by movement of the air. It sets up a new infection by inhalation into the lungs. Airborne infection is thought to be the major route of transmission and pigs are the most infectious animals. The virus may also be acquired by swallowing milk, in meat, by contaminated hay or pasture, and in water. It can also be transmitted on hands, feet, tyres, straw, etc. It can survive for a considerable time in the environment and in other secretions such as milk and semen. Animals may remain carriers for weeks or, in the case of a small number of sheep, for up to nine months.

Dr Ruth Watkins, whose paper 'The 2001 Foot-and-Mouth Outbreak: Arguments Against an Extended Cull' on the use of vaccination against viral diseases is full of useful information, explains that the FMD (foot-and-mouth disease) virus, which belongs to the picornaviridae family, is not 'enveloped' and is very hardy in the environment. 'Virus infections of humans that I am more familiar with, chickenpox, measles and influenza, are all transmitted by aerosol but they are enveloped. Thank goodness they are. There is no human virus that is quite as infectious over long distances such as 20 or 40 miles as has been recorded for FMD.'

The virus is generalised. That is, though it may infect the lung if inhaled, or the gut if swallowed, it replicates locally then spills into the bloodstream and causes a widespread infection in the body of the lymphatic system and organs including muscle. High 'titre' shedding (the strength of the infection in the aerosol particles) ceases as the immune response to the infection occurs: antibodies can be detected on the fourth day of illness, which coincides with a sharp fall in the titre [strength] of the virus.

In 1871 the disease was made notifiable (any cases or suspected cases had to be notified to the relevant authorities) and influential breeders began to lobby for its elimination because they were suffering substantial economic losses as valuable young livestock suffered higher than average mortality. Infertility, occasional abortion and mastitis could also follow infection. Successful efforts at control failed to reduce the spread of the disease, meaning that many farmers experienced movement and marketing restrictions that – in their minds – became inseparably linked to the occurrence of foot-and-mouth disease. The social and economic effects of the existing control measures caused farmers to dread the virus and demand its elimination.

By the 1880s, the battle over whether the disease should be subject to state control was won and the framework of today's controls was put in place. A number of measures were taken therefore: imports of livestock from infected countries were banned; the spread of disease was halted by isolation of infected and contact animals; livestock markets were cancelled and movement restrictions were imposed in infected areas.

Dr Woods writes: 'Whether these measures worked or the disease disappeared for other reasons is difficult to assess, but Britain was remarkably free of foot-and-mouth from 1884 to 1900.' A policy of slaughter was brought in later not least because farmers started to believe it was a severe disease despite the mild nature of its symptoms, simply because of the implications of the very measures undertaken by the state for its control.

During the last serious foot-and-mouth epidemic in the UK, that of 1967–68, slaughter remained the favoured policy of the Ministry, leading to the loss of over 400,000 livestock. This catastrophe was to be totally eclipsed by the events of 2001. The lessons learned during the 1967–68 outbreak applied to far smaller movements of livestock and MAFF's gross underestimation of the number of sheep movements taking place during the weeks prior to 23 February 2001 was to prove a mistake of catastrophic proportions. Professor Hugh Pennington, who had undertaken the investigation into the outbreak of E. coli in Scotland, put it succinctly when he spoke at the *Food Programme* Annual Awards on Radio 4 on 16 June. The department, he said, had finally admitted that it had got the scale of sheep movements wrong. But that is just not good enough. 'It was their business to know!'

This cardinal error led to the initial statements that the outbreak was 'a relatively minor matter' that could speedily be dealt with. To the onlooker it appeared as if everyone in MAFF from the most senior civil servant and government adviser down to the most humble office worker had been, like Thomas the Rhymer, away with the fairies. They had now returned, years later, to discover that the world had changed completely. It would seem that the picture in the mind of the typical

MAFF official was one of little local livestock markets, where farmers in squashed felt hats or checked caps chewed straws, leaned over rails and prodded a handful of heifers or argued over the price of half a dozen sheep. Then a small, nearby abattoir would send a truck to a farm to pick up a few animals to be taken for slaughter ten miles away. But the days of animals being slaughtered in local abattoirs and sold in local markets (as had still been the case in 1967) had gone long ago. In the late 1970s there were 1,350 red meat abattoirs in Britain. By 2001 there were only 339.

Carol Trewin, Farming Editor of the *Western Morning News*, pointed out in her explanation of the rapid spread of the disease on 23 April, that the reduction was due in no small part to 'the remorseless influence of the four major supermarkets [which] has concentrated the meat trade into a handful of super-large abattoirs, most dealing uniquely with only one or two customers of the big four, and perhaps a handful of B-list players from the smaller supermarket chains'. At the same time, the impact of both the BSE crisis and the European Union has led to a raft of new hygiene regulations, all of which have been piled on to the meat-processing companies with huge added costs, which in turn has led to more closures.

So, far from being bought locally, the majority of sheep are now bought and sold in a handful of major markets, not by farmers directly but by dealers, many of whom do not even attend the markets or see the sheep. Sheep are bought, sold, and resold in deals made by mobile phone. By the time the disease was confirmed at Heddon-on-the-Wall, hundreds of sheep on neighbouring farms had already been sold at nearby Hexham market and thousands more at the market in Longtown, a small town in Cumbria only just over the border from Scotland. After which, during the period up to 23 February 2001, hundreds of thousands of sheep had been in contact with infected animals and had then, as Carol Trewin puts it, 'criss-crossed the country travelling thousands of miles and coming into contact with thousands of other animals'.

Sheep from Longtown market, and animals in contact with them, were sent not only to the nearby areas of Dumfries and Galloway, the Borders and Cumbria, but also to Anglesey, mid-Wales, Hereford and

Gloucester, Northampton, Warwickshire, Somerset, Devon and Cornwall, even to Northern Ireland. Nor did it stop there. Both France and Spain still demand the import of live animals rather than carcasses, and one batch of infected sheep, bought by a French stockdealer, was sent to France on 31 January when they must already have been incubating the disease. It usually takes about 14 days to show up, though it can be quicker. However, one of the problems with sheep is that it is often very difficult to tell if they have been infected with the virus. It only becomes apparent when other animals with which they have come into contact succumb to the disease. As far as could be ascertained, the French stockdealer's sheep had not been in direct contact with Longtown sheep, leaving the authorities facing for the first time the nightmare possibility that the virus had been endemic in British sheep flocks from the end of the previous year but had not shown up, and that the sheep had then recovered.

If the frenetic transport of sheep and other animals around the country had come as a surprise to the government and MAFF, it came as a shock to the public in general, not least when they discovered that what they bought in good faith as 'Devon', 'Welsh', 'Hereford', or whatever, lamb, did not actually mean that the lamb was born and bred in Devon, Wales or Hereford. It simply meant that was where the animal spent its last weeks or months before going for slaughter. This is quite legal.

Willie Cleave, a dealer from Devon, bought several thousand sheep from the Longtown market and had a substantial number of them brought down to Devon for fattening. Some he kept on his own land, while others went to various farms in Devon and north Cornwall for the same purpose. On 24 February, just one day after the confirmation of the disease at Heddon-on-the-Wall, foot-and-mouth was confirmed at Cleave's own farm in Highampton. The discovery was to prove the touchpaper that would light the pyres.

MAFF had found it difficult enough to cope with the limited outbreak of swine fever the previous year. It was totally unequipped for what was to come. The main premise of the Northumberland Report was that the way to treat an outbreak was indeed to 'isolate and destroy', but even then it had been conceded that in certain circumstances vaccination should also be considered, and the quality and efficacy of vaccines has improved considerably since then. The whole question of vaccination will be examined in depth later, but even at this stage there were tentative suggestions that it should be given serious consideration and either used as a back-up to slaughter or instead of it. But this idea was firmly stamped on at once by MAFF and the leadership of the NFU, a mindset which has bedevilled the handling of the entire outbreak.

Among the variety of reasons given for their opposition to vaccination, put forward then and afterwards, and amid a plethora of misinformation, was that the EEC would not allow it. This is completely untrue. Had the government adopted vaccination as a strategy it would have been carrying out official European Union scientific policy, for in 1999 the Commission adopted the Strategy for Emergency Vaccination against Foot-and-mouth Disease. The report, by the Commission's own scientific committee on animal health and welfare, states that vaccination can achieve a number of objectives, including a reduction in the numbers of infected animals, the slowing of the spread of the disease, and the reduction of economic loss. The criteria set out for when vaccination would be the best course were: when an epidemic was beginning; a high density of animals in infected areas; proof of widespread animal movements; a suitable vaccine being available; and a steeply rising incidence of cases. All of these criteria were clearly present in Britain by the end of February. However, the government, MAFF and the NFU stuck to their story: that even if they all wished to introduce vaccination the EEC would not allow it.

'We will not use vaccines,' Nick Brown announced in the House of Commons on 28 February. From day one of the crisis, Nick Brown never seemed to be his own man. However tough he might have been with recalcitrant MPs in his role as Chief Whip, as Minister for Agriculture he came across as putty in the hands of his advisers, particularly the two senior civil servants: the Chief Veterinary Officer, Jim Scudamore, and the government's Chief Scientist, Professor David King. Other experts were to be added later, not least Professor Roy Anderson and his fellow-scientists from Imperial College. These latter two were the much-quoted experts whose forecasts of the likely spread of the disease, based solely on computer models, from a team made up of biomathematicians and epidemiologists (basically number-crunchers) would result in the massive contiguous cull policy. Throw into this mix the influential leadership of the NFU, and the fact that the clock was ticking for the announcement of the general election, and a picture emerges of chaos and confusion in which Nick Brown bobbed entirely out of his depth.

It should have been obvious (had a contingency plan been ready) that if the policy of 'isolate and destroy' was to succeed, it had to be implemented with great speed, a fact presumably known to Mr Scudamore who worked on the 1967 outbreak. As matters were to turn out, it was almost certainly already too late for that approach to work when the first case was confirmed in Essex, but even if that had not been the case MAFF simply did not 'do' speed. Its bureaucratic mills ground slowly. Commentators describe it as being obsessed with red tape, rule making, form filling and box ticking. Ian Johnson, of the South-west NFU, described the ministry's handling of events as 'like trying to get a mammoth to run the Grand National'. As a result, infected animals, which should have been culled and disposed of immediately, were left alive for days.

On 27 February local authorities were given emergency powers to close footpaths. Dartmoor National Park had already done so a day earlier, banning all walkers, cyclists and horseriders from the moor, rightly horrified at the prospect that the virus would spread to animals in the wild. Similar measures were taken in Cumbria, especially the Lake District; Cornwall, which closed off the coastal footpath; Dumfries and Galloway, which not only closed its paths and woodland walks but even its lay-bys; and most of Devon. This was the point at which the other massive problem kicked in: tourists, walkers and holidaymakers abandoned the countryside almost entirely, leaving hotels, bed-and-breakfast establishments and a host of other small enterprises without any trade.

One week after the outbreak had been confirmed, Nick Brown made a statement which he was to repeat again and again, sometimes on his own, sometimes as a duet with one or other of his advisers: 'The outbreak is now under control.'

# Chapter 3

# 'Don't Panic!'

The virus raced round the country like wildfire, spreading out most comprehensively from the north and south epicentres of infection, Longtown in Cumbria and the area around Hatherleigh in north Devon. From Longtown it appeared almost at once at a farm near Canonbie, only a few miles away and just over the Scottish border. From there it moved relentlessly up Eskdale, farm by farm. Two of the earliest confirmed cases in Scotland, both originating from purchases by a local dealer, were in the hills close to Moffat. But it also travelled west, jumping the A74 to Gretna where it infected a herd of prize bulls – a lifetime's work gone up in smoke – before racing on along the Solway Firth to the coast, taking in the Isle of Whithorn and Wigton en route. It could not have happened at a worse time for southern Scotland. A poor winter had been followed by a wet spring, then the worst blizzards for years had resulted in many power lines being brought down. People in small towns and villages in Dumfries and Galloway and in Eskdale and Liddesdale had found themselves snowbound and without electricity. However, according to some of those involved, initially at least, the administration in Scotland was to prove more efficient than that in England. The special centre set up to coordinate activity following the Lockerbie disaster (known by its critics as 'the bunker') was opened up again and became the headquarters for the fight against the disease. Almost immediately all lay-bys were cordoned off and disinfectant mats laid not only across lane ends and farm entrances, but across some main roads as well. Children were kept home from school. Walkers were not allowed to walk anywhere except on concrete or tarmac on penalty of a five thousand pound fine. Preparations were immediately put in hand for slaughter.

But even worse than the progress of the disease in southern Scotland was the ferocity with which it hit Cumbria. Cumbria, too, had suffered weeks of poor weather. Farmers there were particularly vulnerable, being so heavily dependent on sheep and they, like sheep farmers everywhere, were approaching the lambing season. Farm after farm fell victim as the virus spread relentlessly along the Duddon Valley, around the slopes of Skiddaw and down Borrowdale.

If the culling process is properly managed, cattle can be rounded up and slaughtered comparatively easily, but sheep are a different matter altogether, many being free-ranging animals scattered over vast areas. Among the various breeds of sheep in Cumbria, the most well-known are the Swaledales (farmed mainly on Alston Moor) and the Herdwicks. Herdwicks, the tough grey and white sheep prized for their wool, come from ancient breeding stock. One suggestion is that they were introduced into the area by Viking settlers in the tenth century, another is that they were already there, part of the flocks of the Cistercian monks of Furness Abbey, who first farmed the Lakeland Dales. As Harry Griffin, one of *The Guardian*'s 'Country Diary' writers, put it on 11 April: 'Either way, they have been indigenous to the area for 1,000 years: stolid, patient animals with sad, but rather noble, white faces; the "little grey sheep" of a hundred books and poems.'

Hard as nails, the Herdwicks live up in the mountains in all weathers, apart from when they are brought down for lambing, dipping and shearing, but that is not what makes them special. As those who knew them were to point out over the succeeding weeks, their most remarkable feature is their homing instinct. They are what are known as 'hefted' or 'heafed' sheep, a term almost unknown outside Cumbria and the Cheviot Hills until the foot-and-mouth epidemic, and of unknown origin. As Harry Griffin explains, on the fell they are 'heafed' (or 'hefted') to their own part of the mountain. 'That is to say they and their families always graze that area and, if taken away for lambing, return there instinctively. Their homing instinct, if once destroyed through mass slaughter, may never be restored.'

Emma Tennant, whose family farms 2,000 acres in Roxburgh and has 1,200 Cheviot sheep, agreed in an article written later for *Country Life* on 19 May. 'Every generation of ewes teaches its lambs to stay on their own ground. Now it would be impossibly expensive and time-consuming to replace a hefted flock since it would take up to a decade of shepherding, day and night. When it became clear that foot-and-mouth had spread from the Solway plain into the Lake District hills to the west, the Pennines to the east and the Cheviots to the north, terror struck the farming families in the hills.'

Those who live in, visit, or simply love the Lake District realised the threat almost at once and were extremely vocal, pointing out that it is the sheep in Duddon Valley, the Coniston and Buttermere fells, in Borrowdale or Wasdale, up to the highest land in England, the Scafells that make Cumbria what it is. 'More than the old dry-stone walls that quarter the fells, the packhorse bridges or the whitewashed farmsteads, the little grey Herdwick sheep typify the Lakeland,' wrote Griffin. Others went further. Without the sheep, the landscape as we have known it for a thousand years, the landscape of William Wordsworth, would soon disappear. Without grazing, bracken will cover the fells and the paths will disappear. Field walls will fall down, meadows turn to swamps.

But it is unlikely this was uppermost in the minds of the authorities in early and mid-March. Herd after herd of cattle were now succumbing to the disease and if sheep were the prime source of infection, then the sheep had to go. However, the prospect of clambering around the Cheviot Hills, the Cumbrian fells and the Pennines, in an effort to round up and slaughter tens of thousands of sheep was, to put it mildly, daunting.

So to the second-worst-hit area, Devon. It was Willie Cleave's sheep, now being fattened in fields in Devon and North Cornwall, which had brought the virus to the region in the first place. As in Cumbria and Scotland, there are sheep that range free on Exmoor and Dartmoor and the prospects for them were dire if they became infected with the disease. But Devon is also cattle country and within no time foot-and-mouth appeared in beef and dairy herds, first in the area round Highampton then quickly spreading beyond. Devon farms are not necessarily very big and soon whole clusters of farms in north Devon succumbed to the disease. First Hatherleigh and Holcombe, then it began creeping towards Dartmoor, to Winkleigh and Okehampton, then on to the moor itself. What still remains at the time of writing, a single case at a farm near Princeton in the middle of Dartmoor, horrified everyone at the thought of the virus spreading first to the sheep on the moor and from them to the moor's wildlife.

Its progress was inexorable. Within days it had reached South Molton and Crediton, Lewdown on the Cornwall–Devon border, Clovelly, Barnstaple, Bideford, Tiverton and on south to Axminster. The pattern of the Devon spread shows exactly what was happening. Confirmed on the first Cleave farm on 24 February, it was confirmed on his second two days later. After that confirmations started coming in first at a rate of two or three a day, then four or five, then more, with no sign of the outbreak even levelling off, let alone declining. Four outbreaks were confirmed in Cornwall and one in Somerset, all directly linked to Cleave sheep.

There were also smaller, but still major clusters in Northumberland, the Brecon Beacons area of Wales (spreading into the Wye Valley) and scattered outbreaks in Anglesey, Herefordshire, Worcestershire, Northamptonshire, Warwickshire, Essex, Kent, the Cotswolds and Gloucestershire. The Gloucestershire outbreak was a matter of great concern as in the Forest of Dean sheep also roam free, covering considerable distances. Like the Herdwicks, the free sheep of the Forest of Dean date back centuries and are part of an age-old way of life that is jealously guarded.

Yet still the pronouncements came that there was no need to panic, that everything was under control, even though half the employees of MAFF appeared, by this time, to be running around like headless chickens. By 8 March the number of cases nationally passed the 100 mark and Chief Vet Jim Scudamore admitted that the outbreak 'might possibly continue for a long time'. The question of vaccination was raised again. Surely, it was suggested, at the very least it could be used as a firebreak to help stop the relentless spread of the disease? Again Nick Brown set his face against it. His advisers had given their verdict: 'The advice to me is very clear. The whole of the industry and veterinary profession are speaking with one voice. They advise against the use of vaccination,' he stated in the House of Commons on 10 March.

This was also not true. Even then there was no such unanimity of opinion. It was not only special interest groups such as the Soil Association and organic farmers that were beginning to push vaccination, but also a substantial minority of other farmers faced with the imminent prospect of losing the livestock they had nurtured for years and, with it, their livelihoods. Papers from a number of scientific experts and veterinarians began to appear in the media and on websites putting forward the case for vaccination. A wide range of experts offered their services and assistance but were rejected, among them a retired vet who had worked on the 1967 outbreak and was considered responsible for preventing the virus spreading to Ireland.

But the die was cast. The only way to eradicate the disease was 'isolate and slaughter'. But, as was pointed out then and many times since, slaughter on the scale needed to implement such a programme, even at that stage, had never been attempted before. Even if MAFF's early calculations on 23 March that there might well eventually be upwards of 918 cases had been right (and it soon proved to be a gross underestimate) it would still have meant the slaughter and disposal of over a million animals. Where were the sheer numbers of vets and trained slaughtermen that would be needed? Who would organise the teams in places as far apart as Cornwall and Dumfries and Galloway? Who would bring in the bulldozers and heavy-lifting gear necessary and where would they come from? Shadow Agriculture Minister, Tim Yeo, demanded that the government call in the army for help, as had happened in 1967, but at this stage his words fell on deaf ears.

Even before the Imperial College team made their apocalyptic predictions, thus initiating the horrendous mass contiguous cull, it is clear that Nick Brown and MAFF were rapidly losing the plot. On 16 March the complete ban on the movement of livestock had been extended indefinitely and that, as we shall see later, was to bring its own appalling problems, problems apparently unforeseen by the authorities that would lead to the deaths of thousands more animals, which were not even infected, on 'welfare grounds'.

The premise behind the 'slaughter and isolate' policy relies on slaughter being carried out within 24 hours of confirmation of the disease and the carcasses being disposed of within 48 hours. From the start this rarely happened. It was a fortunate farmer, if that is the right adjective, whose animals were slaughtered on confirmation and removed the next day. Even in March, when numbers were still (comparatively) low, it could be several days between confirmation and slaughter and several days again before the disposal of the carcasses. Relatively early in the epidemic, a farmer in Eskdale waited for five days with a mass of rotting carcasses outside, attracting rats and other vermin before they were finally burned. It wasn't just the stench that was unbearable, he told a local journalist, but the fact that he couldn't escape the terrible truth of what had happened. Again and again there were no answers to farmers' concerns as to the possible danger of carrion birds and predatory animals carrying the disease elsewhere.

If the setting up of the slaughter policy without ensuring adequate resources were available was mistake two, then lack of forethought as to disposal methods was mistake three. The 1967 Report had posited local burial as the best method and it looks as if, immediately after the first outbreak, this might have been the preferred choice. Except that MAFF then found itself caught in the bind of the previous agricultural catastrophe: BSE. If cattle, some of which might well be over the supposed safe limit in terms of age and/or possibly infected with BSE, were slaughtered and buried, might the waste matter pollute not only nearby pastures, but water supplies as well? The prions that cause BSE are very hard to kill except at high temperatures.

So, in spite of the recommendations of the Northumberland Report, it was decided that perhaps it would be better, after all, to burn the dead animals. But that meant finding great quantities of fuel, including cheap coal, railway sleepers and straw, at breakneck speed. From where? There were no stocks set aside for such an eventuality. Measures were immediately put in hand to get such materials to the areas where they were needed – another logistical nightmare. Getting the fuel and building the pyres to burn the animals was the priority. It seems almost incredible, but at this stage no thought was given to the problems of pollution from such pyres and the consequent possible hazards to public health. That would come later when people finally rose up in protest.

So the first pyres were lit. In the early stages most were quite close to the source of infection, often on the farmer's own land. There are plenty of descriptions of what it was like living close to those early pyres. 'This week,' wrote Fiona Armstrong in the *Sunday Times* on 18 March, 'the stinking balls of smoke hung over Eskdale like a necklace. For those living close to the pyres, there's the soot that blackens cars and homes and the sickening smell of disinfectant and dead animals and paraffin that clings to clothes, no matter how much you wash, and we're all urged to wash regularly, to disinfect and change. A couple in our village have literally been driven out of their homes by smoke from the carcasses.'

During the six weeks from the first outbreak, people on both sides of the Scottish border close to Longtown endured 700 pyres. Residents described how chunks of half-burnt skin from cattle were blown into their homes, the whole ordeal made worse by the use of cheap coal which does not burn sheep properly. MAFF's answer to this was to propose a megapyre, the biggest ever seen, at Longtown. But that, for now, is another story and one that was to prove a turning point in the apparent public acquiescence to what was going on.

It was essential that the government were seen to be doing something, otherwise people might start asking awkward questions so near to a general election. For example, why did they have no proper contingency plan in place for a disease that is endemic in so many parts of the world? Why do we have no effective border controls to monitor meat imports? Why was MAFF so slow to ban animal movements? Why were farmers not briefed at once on the necessary precautions? A local farmer showed the author (on 15 July) a video he had received the previous day from the now renamed DEFRA (Department of the Environment, Food and Rural Affairs) on the need for biosecurity. 'When I asked about this a week or so after the outbreak MAFF told me they were too busy to deal with it!' And what about adequate disinfecting facilities? The difference between what happened here and how the consequent outbreaks were handled in France, Germany and Holland (which has been free of the disease since mid-June) might, if the government were unlucky, actually become an election issue.

The message therefore had to be positive. So, even as the waters closed over their heads, Nick Brown and his advisers could still be heard promising that the epidemic would soon peak, that everything was under control.

# Chapter 4

# The Ticking Clock

Spinners and advisers in the headquarters of the Labour Party in Millbank had been working towards a date of 3 May 2001 for a general election throughout the preceding year, even though the government still had a year of its term to run. The slightest sign of discontent on the backbenches was quelled by emphasising the need for discipline if Labour was to win again. There had already been some blips like the fuel protests and a railway system that had become an international laughing stock. The outbreak of foot-and-mouth was an uneasy reminder that a totally unforeseen issue can arise and spoil the best-laid plans.

Up until March 2001 planning, down to the last detail, had been meticulous. Everything was in hand for the big event. Advertising agencies contracted, scripts for party political broadcasts finalised, poster sites booked for April. MPs joked with journalists about having booked their holidays for May. Well aware of all this, the Conservatives, Liberal Democrats, Scottish Nationalists and Plaid Cymru were doing the same. Extra staff were put on alert, battle buses readied for hire. It was all systems go.

If Tony Blair intended to call a general election for 3 May, the date also fixed for the regular county council elections, he had to make the announcement early in April, then go to the Queen and dissolve parliament. But all of the forward planning in the world could not have foreseen the foot-and-mouth epidemic and Millbank was thrown into turmoil. The result was that throughout March the electorate was treated to endless discussions on radio and television, along with speculation in the press, as to whether or not the election would now take place. However, Blair, his ministers and every other government spokesperson stuck firmly to the pretence that it was only county council elections that were under discussion. What general election?

They were faced with a real dilemma. Any postponement posed problems. The summer months, after June, were ruled out because of holidays – not least the three-month break taken by our legislators. October was a long way off and anything could happen in the meantime: a downturn in the economy perhaps? As for soldiering on to the end of their term of office, few governments have fared well after choosing that course. It does not take a great satirical wit, therefore, to imagine the frantic discussions which must have been going on behind closed doors in 10 Downing Street and the Cabinet Office during the first two weeks of March.

The charade, that only the county council elections were imminent, fooled no one. The Conservative shadow spokesmen demanded a postponement of the general election and made a credible case, but it was all too obvious that they had their own reasons for wanting the election to be put off for as long as possible; reasons that were amply borne out by the subsequent result in June. Liberal Democrats, many either representing or fighting rural constituencies heavily affected by the epidemic, did not want an election either, fully aware how unpopular it would be. In fact it was rapidly becoming clear that there was growing antipathy to the idea of any election at all.

Needless to say, opposition was strongest in the hot spots of the epidemic. Apart from those farms suffering under the A notices which specifically confirm the existence of foot-and-mouth on the premises, many others within the infected areas had now been served with Form D Restriction Notices, which apply when contact is suspected, effectively isolating farmers and their families on their farms. Those within could not get out and unfortunate friends and family stuck outside could not get in. Also, every effort was being made to restrict vehicle movements in vulnerable areas to a minimum. This, therefore, begged the question as to how it would be possible or practicable to hold even council elections when it was virtually impossible for people to get out and vote? This would mean that a whole swathe of the electorate would be disenfranchised, not to mention what would happen in those particular cases where a farmer was actually a candidate.

An array of government spokesmen immediately appeared on news programmes with a cunning plan. The answer? Postal votes. It was being made far easier this time round for people to apply for postal votes, they no longer had to fulfil past criteria such as being vouched for by local worthies. But even if the postman could get to the farms to leave the necessary forms, asked those involved, how was it then proposed that people virtually banned from leaving their properties could get out to post their voting slips? They might just as well go down to a polling station and vote as usual.

But most of the criticism was on a more desperate level. What was the relevance of any kind of election when so many people were suffering so much? For hundreds of farmers, bereft of stock and livelihood, surrounded by pyres and burial pits, and for thousands of other people living nearby, Westminster and its parliament seemed more remote and irrelevant than ever compared with what they saw happening day after day on the ground. The gap in understanding between the metropolitan government and the countryside yawned ever wider and this crisis will leave behind a legacy of distrust that will take a long time to heal.

It seems almost incredible now, but right up until mid-March it does look as if the Prime Minister and his government believed – or wanted to believe – that everything really was under control, as they were constantly being assured by Nick Brown, the Chief Vet and MAFF. Just one more push and it would all be over: over in time to hold that vital May general election. With hindsight it is reminiscent of the generals in the First World War, 'once more over the top, lads, and you'll be home for Christmas'.

On 9 March, Environment Minister Michael Meacher was appointed to head a new Rural Task Force to examine the impact of the crisis on the rural economy. Everything, the government hoped, was going to be all right. Instead, however, this was the point when lobbying began in earnest, not least from a tourist industry rapidly being brought to its knees. Its spokesmen were quick to point out that its earnings for the nation far exceeded those from farming. In 1995, income from farming represented 1.67 per cent of GDP, by 1999 this had fallen to 0.8 per cent. In 1999 (the last figures available) tourism had brought in £16.25 billion. A breakdown of areas favoured by foreign and domestic holidaymakers showed a particularly heavy concentration of visitors in the south-west, centre of one of the heaviest outbreaks, with other epidemic hot spots such as the Lake District, the Wye Valley, the Cotswolds, the Borders and mid-Wales also favoured by visitors.

But in all these areas, the holiday resorts and villages had become ghost towns. Country hotels and bed-and-breakfast accommodation, along with those on the Devon and Cornish coasts stood empty as bookings were cancelled. Only a handful of locals were likely to be found in normally busy tourist pubs and not even there if they were within an infected area. As early as 22 February, NFU President Ben Gill had told people to 'stay away from the countryside' and the Ramblers' Association had called on its members to do the same. Walkers who were determined to walk the paths of the Peak District after notices had gone up prohibiting such activity, received short shrift and nationwide condemnation. In parts of Cornwall it was said that farmers were threatening to greet trespassers with shotguns if necessary. The public heeded the appeals. Anyway, for those whose main pleasure was to walk the hills, moors and woodlands or along coastal paths, there was little point in making the journey to the countryside since they were now firmly closed. As of mid-July it is estimated that the epidemic has cost the tourist industry and the wider rural economy over £15 billion.

Faced with yet another unforeseen problem, government spokesmen, headed by Michael Meacher, toured radio and television studios implausibly assuring the nation, and the world outside, that 'the countryside is open for business', an enraging phrase repeated time after time, which rang increasingly hollow when it was patently obvious that it was not true. Asked on the *Today* programme what exactly this meant in, for example, the Lake District; where exactly tourists were supposed to go and what they could do when they got there, Meacher suggested visiting teashops, walking main roads and doing a little light shopping. But since MAFF and the NFU were still adamantly opposing visits to the countryside unless the journey was essential, the general public could be forgiven for becoming confused by such conflicting advice.

At the very time all this was going on, the government was, unbelievably, in the process of apparently downgrading its own specialist foot-and-mouth laboratory at Pirbright. As a result Dr Paul Kitching, a world expert on viruses such as foot-and-mouth and described by a colleague as 'a gem of veterinary science', had resigned his post and was soon to leave the laboratory for an establishment in Canada. The evidence that Pirbright was now considered of lesser importance was the fact that the advertisement for his successor was for a scientist at a lower grade and on a lower salary scale. Other leading figures from Pirbright were soon to follow Kitching's lead. It seemed very strange, as Magnus Linklater commented in *The Times* on 20 June, 'that in the middle of the world's worst outbreak of foot-and-mouth disease, the very experts needed to chart its progress and to establish Britain as a world leader in animal diseases are departing'.

Once again the question of vaccination was raised, prompting a spate of mixed messages from Nick Brown. 'I cannot rule it out. We may have to retreat to it but it would be a very substantial retreat. By far and the away the best policy is the current one' (19 March); 'Vaccination sounds like the right strategy until you examine what it will really mean. It isn't as easy as it appears' (20 March); 'I want to avoid vaccination if I can but, if it is necessary to bring the disease under control in a speedier way, then of course I have a duty to consider it' (24 March); 'The government is considering whether to use vaccination' (27 March).

At last, in mid-March, the truth of the situation finally trickled through to the Blair government: that in spite of what their Agriculture Minister and Chief Vet continued to say, the situation was not under control. Far from infected animals being slaughtered within 24 hours and then disposed of the next day, the numbers of those waiting days for the slaughtermen and lying around on farms for days afterwards, were growing all the time. And along with concerns expressed ever more vocally and publicly by the farming community, opposition from those living close to pyres or burial pits was growing too.

Eventually it was admitted that there were simply not enough slaughtermen and vets for the immensity of the task facing them. An appeal went out for foreign vets and they were brought in from the European mainland, Russia, even the USA and Australia. But even this was insufficient. Pressure mounted to call in the army; more and more questions were being asked as to why there had been such a delay in doing so. The most commonly held view is that Blair resisted calling in the army for so long because he did not want it to appear, domestically or nationally, that the government had a full-blown crisis on its hands. Blair was about to discover what other prime ministers with control tendencies had found out before him: that it is all very flattering and pleasant to surround yourself with advisers and, in his case, focus groups, who tell you only what you want to hear, but it is also very dangerous.

Whatever the reason for the hiatus, troops finally moved into both Cumbria and Devon during the third week in March, though at first in insufficient numbers. Commanders in both areas immediately stressed that theirs was a logistical and organisational role; they were not going to have direct involvement. This, too, was to prove inaccurate.

Meanwhile, as the number of outbreaks grew ever more rapidly, Chief Scientist David King asked scientists at Imperial College to work out a possible forecast of the extent of the disease, using computer modelling. Presumably he told the Prime Minister the result around 21 March because on the 22 March Blair announced to the country that he was now going to take personal charge of the campaign to eradicate the epidemic. From then on Nick Brown and Chief Vet Jim Scudamore were sidelined in favour of King and the men from Imperial College.

The following day, 23 March, King informed the country that the number-crunchers had come up with a nightmare forecast: that if slaughter and disposal were not speeded up, the number of cases might well reach 4,400 by June. Half of Britain's livestock might well have to go. The way to avoid this, these experts claimed, was to initiate a policy of mass contiguous cull: all stock, however healthy, on farms within a three-kilometre radius of an infected property must, without exception, be slaughtered. It was to prove extremely controversial from the start and there is now, at the time of writing, an enormous body of feeling – and evidence – against it. Also, as at every other step in the fight against the epidemic, very little forethought was given to the obvious implications of what such a holocaust of slaughter would mean. If there were serious delays in the slaughter and disposal of infected animals now, what was going to happen to the position when the numbers were exponentially increased?

Blair now had only just over a week to make up his mind about the calling of the general election. Many in his party, even on the backbenches, wanted him to go ahead. Ignorance of just how serious the situation was out in the countryside was widespread. Others, those with rural constituencies to defend, were more cautious. But at the end of the day it may well have been another overriding factor, one above the legitimate clamour of farmers, the noise of the protesters and the lobbying of the tourist industry, that caused him to draw back. And it was not the reason given by government spokesmen: that now the Prime Minister had taken personal charge of the situation it would be impossible for him to divide his time between that and an election. Instead, it was the fact that night after night for weeks, the viewers of national and local television news programmes had watched, appalled, the stomach-churning pictures coming from the killing fields: the stacked bodies of cows and sheep lying around in great heaps; the sight of carcasses dangling from lifting gear before being flung into wagons like so much rubbish; the pyres of burning cattle, their bodies twisted and bloated, their legs stuck up into the air at bizarre angles. Horrific accounts of what was happening on the ground began to appear in the press. With that as a backdrop, the best spin doctors in the world would be hard pressed to present a Prime Minister in a favourable light, however desperately he might seek re-election. In a memorable Rory Bremner programme Bremner, as Blair, was seen grinning fatuously as he extolled the virtues of his government, while behind him mountains of animals burned on pyres. In the present circumstances, therefore, the Prime Minister apparently felt that all elections should be postponed and an announcement was made during the week beginning 4 April. No firm 'new' date was given for the general election at this point but the obvious choice appeared to be 7 June.

On the day before Devon announced its worst ever toll of confirmed cases over a 24-hour span, Nick Brown paid a flying visit to the county. 'The crisis,' he stated, 'is under control.' Within days Blair formally announced the postponement of all elections and little over a week later MAFF stopped posting essential information about the numbers of livestock involved on its website.

# Chapter 5

# The Killing Fields

It now transpires that on the day the contiguous cull was announced, government policy changed four times in a single afternoon, conducted, according to Anthony Gibson, the NFU's South-west Regional Director, who was interviewed by the author on 27 June, in an atmosphere 'of chaos and confusion'. The policy that finally emerged on that Friday afternoon (23 March) – and which was expanded on by the Prime Minister when he came down to Devon the next day – was that all cloven-footed livestock on contiguous farms, that is, farms with a common boundary, must be slaughtered. This was then slightly modified. All sheep and pigs would be slaughtered automatically, but for cattle there would be a right of appeal. As Gibson points out, however, in practice appeal was a complete waste of time 'because the only appeal they would allow was if your farm didn't actually adjoin another farm. Even if your cattle were miles away from infected stock, but you had a 500-metre hedge on the very boundary of your land which touched the far boundary of an infected farm, your stock still had to go. That caused great anguish.'

In Cumbria, with its vast herds of sheep and where the three-kilometre cull was strictly observed from the start, the slaughterers made the sheep a priority, leaving cattle alone for the time being. This saved at least some of those animals which had remained for 21 days without sign of infection; although not as many as the disgruntled Devon farmers, who felt that the cattle farmers in Cumbria had it easier than them, believe. But sheep flock after sheep flock were culled in rapid succession. In one district alone, that of Bolton by Bowland, 47 farms were 'taken out' in less than a month, leaving only a single small flock of sheep still alive, and six farms still unculled.

In those early days there were numerous stories in the national press of what was happening in Cumbria, but a first-hand account of what it was like on the ground there comes from farmer Richard Mawdsley, who kept a brief journal of his own experiences and those of some of his friends and neighbours:

*Saturday*
Breakfast and an unhappy Joan, wife of a farmer who is overwintering my in-calf cows, calls to say one of their milking cows has sores on her mouth and a temperature. Ministry vet called in. Teatime she calls again to say the vet is on farm and foot-and-mouth confirmed, despite every possible precaution: children taken from school, Joan giving up her part-time job as a care-assistant, disinfectant everywhere, yard closed to all visitors all the time since the first outbreak three weeks ago. Auctioneer from Cockermouth tracked down on his night off to carry out valuation of cattle on Sunday.

*Sunday*
Cattle valued pending slaughter. Ministry vet was good, kindly and efficient, one of those drafted in from overseas, in this case from the USA. He told Joan and Nick that in the States they realised two weeks ago that we had an epidemic on our hands and were shocked to hear the government saying that everything was under control.

When Mawdsley's friend told the vet that local farmers were questioning the wisdom of building huge funeral pyres and commenting on the number of new cases downwind of them, the vet ended up admitting that he too questioned the wisdom of burning.

> Afternoon, went to see a neighbour, James, whose brother's cattle had been destroyed on Saturday. Wished I hadn't. There are no adequate words to describe the horrors on the Solway plain, north of Wigton. In the village of Little Bampton, all the livestock has now gone. On one farm next door to James's brother, slaughter was promised the day after diagnosis and an A notice served on the farmer, virtually confining him to his yard. Four days later the slaughtermen arrived, during which time cattle in another yard down the road couldn't be fed. After the killing, the bodies were spread out and left. By the time they were collected, some of the bodies had gassed up and exploded. Some were so distended that they had become jammed against one another. Others exploded when they were picked up. The stench was overpowering. Blood and filth were running from the yard down the road and getting into the drains. Right across the road on another farm was a heap of sheep carcasses awaiting disposal. Crows were commuting between the dead bodies and feed troughs. Elsewhere, a heap of sheep carcasses lay, torn and scattered by foxes, before anyone got around to collecting them.

The inevitable happened and foot-and-mouth reached James's brother. 'The vet, another drafted in from overseas, was most outspoken. Burning infected carcasses, he said, spreads disease. No cook puts the meat in the oven then lights the fire! The bodies are put on the pyres which are then lit. As the heat builds up, the bodies first give off water vapour which carries the virus. Only when the fire is burning fiercely enough is the virus killed.'

Mawdsley then returns to the story of his own infected cattle. The friend who had been overwintering them had been assured of an on-farm burial and a digger was laid on for the Tuesday, two days after the slaughter, but it was held up on another job. 'Wednesday morning, the contract boss was actually in the yard when he was told a ban had been imposed on all on-farm burials for fear of BSE getting into the ground water.' The farmer was then told he would have to wait ten days until incineration could be arranged, or four days for a mass burn arranged by the army on a site that had still to be found. However, after a further two days, another contractor with a digger arrived, stripped off the topsoil and began work on a deep hole, at which point his boss arrived and stopped him. Someone had failed to pass on the new instructions. 'The hole was filled in and levelled out. Then the mass burn was cancelled. Finally some, but not all, of the carcasses were removed.'

Finally Mawdsley and his friend were told that trucks were arriving to collect the rest of the carcasses. To make it easier for them to get into the farmyard, 15 tons of stone were ordered to firm up the roadside verge. But after the first two loads had arrived, a further 60 tons were tipped on the verges and yet another digger had to be called in to level it out. Finally, after eight days he writes:

> the stinking, bloated remains of the herd have gone. One thing has improved. The rear doors of the trucks were sealed with expanding foam before loading and each one was followed by a car to ensure that no leakage would pass unnoticed. In the small market town of Wigton on the North Cumbrian plain, the stink of rotting flesh could be smelled in the streets at the weekend. In Little Salkeld, a village a few miles East of Penrith, people can't sleep for the stench that fills the houses. There are still heaps of bodies in yards and in the fields all through the east and north of the county. In village after village there are no longer any animals to be seen or heard.

A neighbour of Mawdsley saw his 200 sheep and 8 in-calf cows shot one Sunday morning. He had already lost another 600 sheep which were away wintering and were caught up in another outbreak. 'Another flock of hefted sheep gone.'

Two weeks after Mawdsley lost the cattle on his friend's farm, a Ministry vet arrived to check his surviving livestock. To his surprise he discovered the vet was a Spaniard 'whose command of English was barely adequate. He was astonished to be told he was my first contact in two weeks. "You must realise that none of you are very important," he said. "All they can think about is their election." Then added, "And you won't be too surprised when I say that the figures you see on the TV in the evening don't tally with the ones we see coming into the office." His English was good enough for that.' On 5 May Mawdsley had a call from MAFF to say that as he owned the two fields across the road from an infected farm, the sheep he had in them would have to go. He notes that he was fortunate they were sufficiently far away to save his immediate livestock. 'Last week nearly a thousand head of dairy cattle (all the cows in milk, and young stock) were slaughtered in and around Udale village. There are now only three flocks of fell sheep left in the whole of Udale parish. And still the plague creeps nearer.'

What enraged Cumbrian farmers, and farmers elsewhere, was the fact that the rules kept changing. Sometimes they received contradictory advice from two people on the same day in the same office. First they were told there would be no vaccination, then that there might be, but only as a stopgap measure to be followed by a cull. Next they heard from Nick Brown's office that he had never had anything against vaccination and that he had not said it would have to be followed by slaughter. 'Then,' said Mawdsley, 'we were told that we weren't to get it because "the farmers" don't want it.' Yet in a survey carried out in Cumbria 85 per cent of farmers were in favour of vaccination, not least as a measure to save what remained of the hefted flocks of sheep. 'The Ministry says it will do everything in its power to try and save the hefted hill flocks and some of the other rare breeds. Having read two of their papers I've come to the conclusion that they have no real understanding of the subject.'

In Devon, as in Cumbria, what was actually taking place on the ground bore precious little resemblance to what was being purveyed by politicians, MAFF and the variety of experts hauled in by the media to give their comments. It might well have been happening on another planet.

The toll of infected animals in Devon was rising so rapidly that by the time the three-kilometre contiguous cull was announced and the army was finally sent in, it had already reached 56 confirmed cases. The delay between confirmation and slaughter was already well behind the 24-hour deadline, and the wait for disposal afterwards was even longer. The contiguous cull was about to make matters far, far worse.

Within a week of the disease arriving in Devon it became obvious, says Gibson, that MAFF vets were not coping. 'They couldn't cope and never would in the circumstances. They were just not equipped to build vast pyres, composed of hundreds of tons of material which had to be transported down a back lane to some farm in deepest Devon. That was part of the legacy of the 1967 outbreak. Planning was based on farm size then, when farms were only a third of the size of what they are now. In those days you could dig a pit on site, bury the animals and cover them with quicklime. The sort of numbers involved in Willie Cleave's place – a thousand head of cattle and 5,000 sheep – required a major civil engineering job. MAFF vets had no equipment to make that happen. Yet when we told them this and urged them to send in the army, they got all uppity and took it as a slur. The message they were sending to their masters in London was "don't worry, everything's all right, we're coping". Meanwhile we were sending messages saying it's not true, everything here's bloody chaotic, it's taking days to cull animals, and farmers were ringing in asking why MAFF's Exeter office was saying everything was fine when it wasn't. In the end the poor chap in their regional office was like a cork tossed about in a storm. I felt so sorry for him. As for Nick Brown, he was badly let down by his officials and MAFF.'

MAFF's contingency plan for Devon had posited 50 outbreaks at the most, but 1 April the toll had already passed 100.

But Devon had an added problem. The period mid-March to mid-April 2001 was one of the wettest on record. Sheep, brought down for lambing, were now outside in fields that soon became quagmires. The obvious course would be to bring the animals into the dry, or at least on to drier pasture, but farmers covered by D restrictions were unable to move them to safety even from the opposite side of a lane or road; they had to stay where they were; nor were farmers allowed to take food to them. Conditions were soon appalling.

Lambing began. Some sheep died, giving birth in a landscape that looked like the Somme battlefield. Others were seen staggering around with half-born lambs hanging out of them. Lambs were born and drowned or suffocated in the all-pervading mud. Crows were seen pecking out their eyes as they lay dying. The situation was little better where farmers and their families could get to the sheep, as they could not move them anywhere else. West Country television viewers were faced with ever more harrowing images night after night. RSPCA inspectors fulminated against a system which caused such suffering, rightly pointing out that under normal circumstances any farmer in the country who allowed his stock to live, often unfed, in such appalling and inhuman conditions would find himself in court, heavily fined and forbidden to keep animals ever again. Eventually in several instances they were unable to stand by and watch any longer – in one case inspectors brought in bales of straw to try and sop up some of the mud in a field littered with dead and dying newborn lambs. Surely, the farmers asked, they could at least be put out of their suffering?

Perhaps this is a relevant time to raise a matter which has exercised a good many people since. Where were the animal rights activists when all this was going on? Where were the protests, the marches, the high-profile stunts? Some of those who enquired were told that this was not a matter for the animal rights lobby as so many millions of animals went for slaughter each year anyway, so what was the difference? And we all know about the ghastly conditions that exist on battery farms and intensive stock rearing units. On the whole there has been complete silence from animal rights activists. Even for those who have hitherto had a great deal of sympathy for their cause, the thought must cross their minds that perhaps protests about dying animals in remote fields and mass slaughter were not as politically interesting or effective as picketing say, Huntingdon Life Sciences, a picket which went on throughout the contiguous cull.

The case was different, but also desperate, for pigs. Pig farmers with fattening units found their animals getting steadily bigger while the amount of room they had in which to move remained the same. Again, surely it should have been possible, if there was no alternative, for them too to be slaughtered on welfare grounds?

Belatedly as ever, towards the end of March, Nick Brown

announced an animal welfare disposal scheme designed, he said, to relieve the severe problems being caused by the weather and the movement restrictions. It was welcomed with relief. It would, he insisted, commence within days of his announcement. But, once again, this simply did not happen. MAFF, still blanketed in bureaucracy and smothered in its own red tape, simply was unable to get it together. Two weeks later Nick Brown had to admit that out of 1.6 million animals identified as being in desperate need of culling for welfare reasons, only 48,000 had actually been dealt with under the new scheme.

In the limited number of places where movement restrictions were lifted on welfare grounds, farmers wanting to move their animals more than 500 metres had to apply for a special licence for each movement and have their vehicles inspected to ensure they met with a new set of standards. Then each vehicle had to be taken away to an official disinfectant point, which could be miles away (thus defeating the object of the exercise) and disinfected under the eyes of MAFF officials.

By the time Blair 'took control' the number of infected farms in Devon had risen to well over a hundred and involved tens of thousands of animals. It does not take a mathematical genius or a number-cruncher from Imperial College to realise what was bound to happen when you added both the contiguous cull toll, and the animals waiting to be culled on welfare grounds, to the original estimates of how many animals would need to be slaughtered. What is also appalling is that we will probably never ever know the true toll of the dead for, from the start, the numbers of farms and animals slaughtered within the contiguous cull area were never accurately collated, and at the time of writing are no longer even being counted officially. Towards the end of the second week in May, MAFF slaughtermen and vets in Devon realised that the task before them was virtually insurmountable for they were now faced, behind as they were with slaughter and disposal on the infected farms, with having to deal with several hundred more farms and tens of thousands more animals.

As the massive slaughter got under way over the Easter weekend, 'the bloodbath', described by Gibson, the first voices were already being raised against a policy based on computer modelling rather than practical common sense. The slaughter policy was, however, enforced with particular determination by MAFF, especially over that Easter weekend. MAFF slaughter teams arrived on farms with very little warning, often very early in the morning. Nor did the examples of healthy animals being culled before infected herds do much to reassure critics. With an awful inevitability the carcass mountain began to grow. First 22,000 dead animals left lying around for days, then 150,000, until the tally reached a staggering 176,000 rotting carcasses. Many animals were culled right next to the farmhouses and buildings. Bloated carcasses of cows lay piled up in the yards outside dairies or immediately outside the farm gate. Great hills of dead sheep were mounded against walls beside roads and lanes, clearly visible to those of us unfortunate enough to live amongst it. The stench was intolerable. One farmer had to live beside his rotting pedigree dairy herd for a fortnight. Again the resulting pictures were shown on local television.

So, not only did farmers have to stand by and watch the total destruction of their life's work, herds and flocks built up over 20 or 30 years killed before their eyes, but they were then forced to watch day after day as the livestock on which they had expended so much care rotted in front of them; animals which all too often were quite free of the disease. Some of those interviewed on television broke down and wept. Even animals that had been kept almost as pets had to go.

What anyone living in one of the affected areas will remember for the rest of their lives is the smell that pervaded everything. Driving through such areas it permeated the car even with the windows and sunroof tight shut. The whole of north Devon stank. 'It was,' says Gibson, 'one giant, open-air charnel house.' The authorities had 'roared' into the mass cull policy without a notion of how they were going to deal with all the remains. Their obsession was with the 24–48-hour culling process. 'They were completely focused on that and so the carcasses continued to pile up.'

While the situation was at its most acute in Cumbria and Devon, other areas were not faring well either. By 7 April Wales had notched up 48 cases. The biggest concentration was in Powys, followed by Anglesey with 13, and a scatter of others in Monmouthshire and Conwy. Visitors to Trecastle in the Brecon Beacons just before Easter were greeted by large placards declaring that they were now entering

'Death Valley'. By 25 April the number of animals culled in Wales had risen, according to the National Assembly, to 300,000, with a further 200,000 awaiting slaughter.

Rhodri Hornung, political correspondent of the *Western Mail*, put it succinctly in a piece written on 24 April:

> There is no better way to describe it – the battle against foot-and-mouth has been warlike. Death, destruction, slaughter, culling, trenches, pyres, army involvement, operation centres – all the images of a full-blown war have been present . . . Even if figures show a dramatic tailing-off of the disease, we can ill afford to be complacent, as Britain's Chief Veterinary Officer, Jim Scudamore, no doubt knows to his cost. His remarkable statement of early March was definitely a premature declaration of victory. 'The spread has already taken place,' he said, confidently. 'What we are doing now is mopping it up and following it up.' How wrong could he have been? At that time there were just 32 confirmed cases across the UK, three of which were in Wales. By the beginning of this week the figure was approaching 1,500, with 75 cases in Wales.

By Easter weekend a number of websites were up and operating in an attempt to keep people informed as to what really was going on and that much of it was deeply unpleasant. One of these was Sheepdrove.com, set up by ex-publisher and organic farmer Peter Kindersley and his wife, and others include Farmerstalking.com and, putting its own views on-line, Cullmaff.com. While it is impossible to check out all the stories coming into these websites, and taking into account the naturally highly emotive state of some of those who had witnessed upsetting events, there are a number of notable instances which were cited by several people, each of whom logged them on to websites. Two examples will suffice.

First the story of the farm next door to 'Rabbie Burns farm', in Dumfries and Galloway. It belonged to Colin Stokes, a wool weaver and organic farmer who had been developing a new breed of sheep he called Braydons, for 15 years . They were, as one witness put it, not only unique, but also famed for the quality of their coloured wool. There were 200 of them, they were quite healthy and they were on the very edge of the three-kilometre cull area. The onlooker was deeply disturbed by what he witnessed. 'An attractive blonde was bending over a small pen of chocolate to beige-coloured lambs, all bleating piteously, as she systematically jabbed them with a long and lethal needle, and left them to stagger aimlessly until they collapsed beside their fellows. A growing pile of limp, flattened little figures lay beside the pen. Two other pens were crammed so tightly with sheep that it was not immediately obvious the occupants of the furthest pen were already dead.'

At about the same time the *Forest of Dean and Wye Valley Review* ran the story of a pet cow called Moo:

> A Forest animal owner and the local community are outraged after Moo, the cow who thought she was a horse, was slaughtered in front of a pub owner and diners last Sunday lunchtime. Moo's owner, Lesley Wogan, was devastated. She had been battling with MAFF officials to try and save the cow which was in a paddock with a pregnant mare opposite a local pub. Eight MAFF officials turned up. The wife of the pub landlord, Yvonne Penketham, watched in disbelief as the cow was killed by fatal injection, then removed from the paddock by a mobile crane in front of diners eating their lunches. People were in tears and unable to continue eating and Mr Penketham did not charge them. The cow had spent a great deal of time in the paddock beside the pub and was a pet to local children and she always came to the gate when they called her.
>
> Mr Penketham called MAFF the next day to tell them how upset he was but they said they had had to act quickly to avoid protestors and they had refused to wait while the horse was removed first. Mrs Wogan had asked for her cow to be tested for foot-and-mouth but was told it was too expensive.

On top of everything else they had messed up, MAFF's public relations were abysmal. There seemed to be no comprehension of how their policies and actions were going down with the public in general. Perhaps they did not think it mattered or that people would assume that, unpleasant as it was, this was an emergency and drastic action was necessary. Possibly they thought the images of slaughter could not get any worse. But that was before the pictures of the pyres and burial pits appeared on television news programmes and the front pages of every newspaper in the country.

# Chapter 6

# Pyres and Pits

The relief of having finally decided to postpone the general election did not last long. The clock was soon ticking again and, with Nick Brown pushed to one side, the spotlight now shone directly on Blair, who had very publicly put himself in charge of the crisis. The message was still that everything was under control. The Chief Vet claimed the proof of this was that the number of cases had now dropped to only 20 a day when, at its worst, they were coming in at 30 or 40. However, Lieutenant-Colonel Barry Hawgood, spokesman for the army's operation in Cumbria, was not so sure. Quoted in *The Guardian* on 24 April he said: 'It's a very dangerous thing to say that the epidemic is now under control. So far as Cumbria was concerned, we're confident that we are in control of the situation as it is now, but if we have a flare-up, a new outbreak that we didn't anticipate, it could lead our resources to a level we do not wish to use.'

This view was later supported by Cumbrian farmer Harold Armstrong who stated in an interview with the author on 13 July: 'Around here we're now on target, slaughtering animals within ten days of reporting the infection, but that doesn't mean it's under control. Nationally the number of farms infected is falling, but the percentage of cases in some areas like Cumbria is rising. From here I can see twelve farms around me that have been infected, and only four that are left.

Two-thirds of farms have been infected and those that are left are on their own, so if they're saying the number is falling, that's because there aren't many farms left.'

By now Blair was adamant that there had to be seen to be a real breakthrough by the end of April. If there's any one thing that enrages Tony Blair more than another, claims Anthony Gibson, 'it is perceived incompetence. And the one thing MAFF was producing in vast quantities was just that: incompetence. He was absolutely determined to sweep it all to one side and get it sorted before the final date for calling the general election.'

Given the goal towards which his entire administration had been aiming for so long, this had to be early May for a June election. There could be no further delays. The outbreak had to be seen to be nearly over, whatever it took, however terrible the cost.

While, given the state of the opposition, there was very little risk that the government would not be returned to power, Blair knew there were an awful lot of disenchanted voters out there. Labour, 'New' or 'old', had never attracted the farming lobby, of course, but now there were farmers who found themselves not only without an income of any kind, but unable – due to the restriction notices – even to get into the nearest town to try and sort out a bank loan or see if they might be eligible for benefit. The situation in the west of England soon became so bad that the *Western Morning News* launched its 'Green Wellie' fund. Green wellington boots appeared in shops, pubs and other public places into which customers were invited to throw their loose change.

Still those dreadful images were appearing night after night on television screens, for, although the army was now working flat out, the carcass mountain continued to grow. Therefore even more drastic measures had to be taken. Those living in infected areas were all too familiar with the sight and smell of local pyres on nearby farms, indeed for those of us living in the West Country they were impossible to miss. Some farm burials had reluctantly been allowed, but this had fuelled immediate expressions of anxiety in the neighbouring communities about a possible re-emergence of BSE from cattle, leading to more cases of its human form CJD. Support for burial as the answer had not been helped when in several instances it proved necessary to dig the dead livestock up again after what was delicately and euphemistically described as 'waste matter' had bubbled up to the surface from the burial pits.

Finally, government experts and MAFF came up with the solution: mega-pyres and mega-pits, able to deal quickly and efficiently with vast numbers of dead livestock from a wide area. To those sitting in Westminster, hundreds of miles away from the centres of infection, it seemed an excellent idea. Clear all the backlog of carcasses away from the farms and destroy them all in one go; that would solve everything. A few voices were raised to the effect that it might not go down too well as a public relations exercise, but it was pointed out that there had already been a number of pyres built which were substantially bigger than the average without too much protest. To which the critics countered that was all very well, but the result of the exercise had been the dramatic pictures on the front pages of newspapers of what appeared to be one of the circles of Hell from Dante's *Inferno*: long lines of fire set against the night sky, in which the silhouettes of hundreds of twisted corpses could be seen. These images too had duly appeared on television.

As for burial pits, in spite of the hiccups where dead animals had to be disinterred, it was decided that if they were big enough, and advertised as being properly monitored with regard to the 'waste matter' involved, then hopefully those living in the vicinity would accept them as a necessity, given the gravity of the situation. So the search was on for suitable places and sites were quickly selected in Cumbria, Wales and Devon. The sites chosen for the mega-pyres were a disused airfield, Hallburn, in Cumbria; part of an army firing range in the middle of the Brecon Beacons, known as the Epynt; and a piece of farmland next to the town of Holsworthy in Devon. A site for a burial pit in Devon was also earmarked, at Ashe Moor.

Again, looked at from the safe distance of Westminster and Whitehall, the sites seemed very suitable: all miles off the beaten track and away from any large conurbations. But while the designated areas might look relatively uninhabited to the casual map-reader sitting in an office in London, it was not the case in practice. It seems that the authorities did not realise that hundreds of people actually lived and worked in the designated areas, people who would soon find themselves living cheek by jowl with the mega-pyres and mass graves. So the protests began.

One of the first to impinge on the general consciousness was that in Cumbria.

On 7 April work began on constructing a 500-metre-long pyre at Hallburn airfield, and such was the speed of the operation that by the next day the new access road required was already in place. There had been many complaints locally about the cheap coal used on smaller pyres which caused unpleasant smoke, those living nearby complained that it simply was not good enough to burn sheep properly and pyres smouldered on for days, even weeks. The response of Ian Pearson, Regional Disposal Officer for MAFF, was that in view of this, special low-sulphur coal was being imported from Colombia and that a grid of steel girders would be built under the pyre to ensure a proper up-draft. Pearson told enquirers that this would result in 'a new-technology burn, a faster, efficient, higher-temperature burn'.

But people living both sides of the Border were singularly unimpressed. They had simply had enough, far more than enough. When they were informed, as the fires were lit, that the pyre was now ready to burn up to 3,000 carcasses a day – for a year if necessary – anxiety turned into enraged protest, stoked by the knowledge that wagon-loads of dead animals were even now on their way from all over northern Britain. During the night of Thursday, 12 April alone, 83 of the 'death wagons' had been seen making their way to Hallburn.

Matthew Engel, writing in *The Guardian* on 14 April, put the position well:

> This was to be Britain's biggest animal crematorium. Its rapid construction was, at last, the display of fast-moving efficiency for which the nation had been waiting. It was probably a re-enactment of the kind of wartime effort that built the airfield in the first place. Unfortunately this was under the control of MAFF, an organisation so hapless that if it began selling umbrellas, the rain would cease.

Yesterday it was obliged to close the site for at least a week. It might never reopen. And if it does, unless there is a dramatic change of mood, it may have to do so over the prone bodies of local residents. With amazing crassness officials have chosen Hallburn, less than a mile from the centre of Longtown, the small town outside Carlisle which has attained a sort of infamy as the foot-and-mouth capital of the world.

He pointed out that Longtown had been surrounded, sometimes covered, in choking smoke for weeks and that on the six-mile drive between the town and Carlisle not an animal remained alive. The first view of the town for those driving north was a pyre of sheep at Howend Farm, still burning after five weeks.

The people struck back and a public meeting was called in Longtown community hall. One of the features of such protests, whether on the Borders, in Devon or in Wales, has been the patronising complacency with which the authorities have greeted their critics. The 'we-know-best' approach to those hayseeds, eco-freaks, troublemakers, stroppy farmers, sandalled organic food eaters, and little old ladies that inhabit rural Britain. The badly briefed MAFF representatives and local authority officials who faced their hostile audience in Longtown were therefore in for a shock. 'Nothing,' wrote Engel, 'could have prepared them for the fury that would follow. It was, in that sense, a microcosm of the way the epidemic has overwhelmed thousands of other well-meaning public servants, up to and including the minister of agriculture himself.' It all got off to a bad start when Peter Tiplady, Director of Public Health for North Cumbria, made the over-smug announcement that the chances of BSE spreading from the pyres was 'one in a million'. 'You don't give a shit,' was the prompt response. 'You're just happy to come to Longtown and set fire to it.' The complainant sat down to wild applause. The protestors then put up James Bell. Mr Bell is a research and technology director for AMAT, a company devoted to advanced environmental technology which is recognised as a world leader in carbon chain degradation. He is also the son of the retired Longtown vet who, during the 1967–68 epidemic, prevented the disease spreading to Ireland and whose expertise had been rejected. The nature of Mr Bell's work means that he has to have considerable knowledge of emissions of hydrocarbons: 'I have to have in order to obviate them. I need to know what is emitted from burning tyres, plastics, any waste.' He is a forthright person who does not suffer fools gladly. He told the panel that he had 13 questions to which he would require answers. When he got as far as 'polycyclic aromatic hydrocarbons', it was clear so far as the panel were concerned, that he might as well have been speaking in Aramaic. The decision of the chairman of the panel to try and wrestle the microphone from him at this point proved, noted Engel, 'to be a mistake . . .'

Bell persevered and soon the reference numbers of two little-known EEC regulations were brought to the attention of MAFF, the EU R45 Requirements instanced in 75/552/EEC and 67/548/EEC, both of which deal with pollution. Pollution? What pollution? It was only smoke, after all. It seems hard to believe but up until then no one had thought it necessary to monitor what was actually present in the smoke given off by the pyres.

Bell told the panel that he had calculated that the smoke would contain eight carcinogens, various dioxins, and possibly mustard gas and, even with the special low-sulphur coal, 9,000 tons of sulphur dioxide. No one countered his claims, or answered his obvious question as to whether or not the smoke itself might contain the foot-and-mouth virus. If that did prove to be the case, then the wind would carry the flames either over the town or towards the Lyne Valley, the one part of the area still disease-free and with 11 farms remaining uncontaminated. The Lyne Valley was also the route designated for the wagons carrying the remains. John Armstrong, a farmer who had played a prominent role in the protest, said that they were supposed to be sealed 'but I know that at least one of them has had blood dripping'. His neighbour complained that the wagons were themselves a hazard. 'The drivers have got no consideration whatever. They're going so fast along these narrow roads they're going to kill someone.'

A woman intervened to say that her garden and windows 'were smothered with globules from the burning. It's like resin. Even a woman knows that what goes up must come down.' Finally, the area's

NFU secretary, Nick Utting, stood up and told Ian Pearson, 'I'm afraid, sir, that you cannot expect Longtown to take any more. The people have had enough.' Faced with such overwhelming determination and opposition, Pearson agreed to recommend to the relevant authorities that the pyre be closed at once. At the time of writing it remains closed. In all it burned 3,000 carcasses and has cost the taxpayer over £2 million.

The people of the Epynt were not so fortunate. 'Is this the longest burning pyre site in the UK?' asked the editor of the website Epynt-disaster.co.uk on 19 June. It had then been burning since 11.58 a.m. on 5 April. People living around the Epynt, as the *Western Mail* explained, have probably had as much history as they can take, because over the last century most of it has been bad. The decision of the government, backed by the National Assembly of Wales, to allow the army range to become the main site for the disposal of slaughtered livestock merely proved to be the last straw.

Living within 600 yards of the pyre site in a village called Llandeiloi'r Fan, internationally renowned theatre director Michael Bogdanov told BBC Wales on 10 May what he thought about it, what local people thought about it, and how events had conspired to provoke such rancour. In 1939 people living in the Elan Valley, Trewyn and the Epynt were given only days to get off the mountain before the army took it over. Though compensated, said Bogdanov, there still remains sadness among the displaced and their descendants over what they lost. Next, in the 1960s, came plans to flood a local valley to provide water for England, but these were successfully fought off. In one way the Epynt protestors were lucky to have such a high profile champion – Bogdanov is hardly a shrinking violet when it comes to publicity – but in this case his profile was to be of no avail.

The pyre at Epynt went ahead despite the valiant attempts of the local community to prevent it. On 1 April there was a mass demonstration as protestors blocked the road at Sennybridge in an attempt to halt lorries carrying straw and wood for the pyre, while others blockaded the entrance to the site itself. Both sets of protestors were removed by police. A Dyfed Powys police spokesman described the protests as 'good-natured' and the protestors as 'mainly local people'. Glyn Powell, deputy president of the NFU in Wales, who farms in the area, told the local media that the demonstration would continue throughout that weekend. 'It is as if these people have a death wish for the farming industry in general,' he said. 'The protest is a way of showing solidarity with the community.'

The initial good humour was, however, to be marred by a dramatic incident when on 2 April, following a day of clashes between protestors and police, a protestor stole a bulldozer and drove it at speed through a security cordon. It collided with an army checkpoint and a number of military vehicles, before crushing a police van in which two officers were sitting. One jumped to safety but the other was trapped in the wreckage and it took firefighters two hours to free him. He was later taken to hospital in Abergavenny with leg injuries.

According to the Chief Constable of Powys, Terence Grange, in a statement made later, more than 300 protestors had been throwing 'projectiles' at contractors' vehicles and trying to block access to the site on the Saturday. 'They were hiding in the bushes and jumping out in front of trucks. The roads were blocked with JCBs and bales of hay and they were putting oil on the road to make it impossible for the lorries to get up the hill.' But by the small hours of the morning, only about 50 or 60 protestors remained and they were being 'escorted' down the road by police officers 'when this enormous vehicle came down the hill and drove over one of my vans. It was squashed with the driver trapped inside. Everybody was amazed he was alive. The officer is badly bruised and traumatised . . . this is an absolute horror. It must have been the most terrifying thing to go through. It was not just officers; everybody there was fleeing for their lives.'

The JCB was later abandoned on the nearby A40 and a man arrested at his own house. Three other local men were also detained by the army on their ranges and later arrested by police officers on suspicion of the theft of the machine.

The Epynt pyre, the protestors were told originally, was intended to burn for only two to three days. Preliminary assessment of carcass disposal sites in Mynydd Epynt, prepared for the Environment Agency on 4 April, explained that to burn 300 cows it would take 175 tonnes of coal, 380 railway sleepers, 250 pallets, four tonnes of straw and 500 gallons of diesel. What, the protestors now ask, is the estimate then of an 80-days-plus burn? How much material has been used? How much dioxin, mutagen, carbon monoxide, phosgene, acidic acids and carcinogens, including formaldehyde have been released into the atmosphere? How many BSE prions, how much foot-and-mouth virus, salmonella and E. coli? Not to mention possible arsenic from railway sleepers and vapour from the plastic covers which had been on the lorries. At its worst, the smell of burning flesh on the Epynt was reported 40 miles away by a farmer at Boncath, near Cardigan. He rang the Epynt Action Group to say he could smell it all one day when he was ploughing on high ground and the wind was coming from the direction of Llandovery.

But the pyre was not the end of it, for the people of the Epynt soon learned that they were to host burial pits as well. By 7 April the police were back in force. Trecastle village was described as 'teeming with police officers' sent in to deal with protestors. Pubs, bed-and-breakfast establishments and the local primary school stood empty, but worst of all was the terrible stench from the lorries carrying the carcasses to the mass grave. Lewis Chamberlain, owner of the Castle Coaching Inn Hotel, said it was hardly surprising tourists stayed away from what is usually a popular area. 'Let's face it. Who wants to go to Cambodia for their Easter break? Trecastle is seen as the front-line in the battle between the environmental lobby and the authorities.'

Councillor Meredydd Jones told the *Western Mail* it was like living in a police state. 'The number of police officers is unbelievable. They've come from as far away as West Mercia. The local council, Powys County Council and the MP weren't told anything about it. Nobody knew anything until Thursday, and they began digging the pit on the Friday.' Villagers and farmers alike were extremely concerned that foot-and-mouth would break out in the area. They were also desperately worried that seepage from the carcasses would leak into the local rivers and contaminate their water supply. But their biggest fear was that the Epynt was to become the dumping ground for the whole of Wales.

The ensuing weeks would show that the protestors had a great deal to be concerned about. Soon infected carcasses from all over the county were being taken to the Epynt, while some of those already buried in the notorious pit had to be dug up again weeks later and were then taken back to their starting point for incineration rather than using the pyre close by. Witnesses describe the wagons as a horrible sight, dripping stinking fluids and with appendages of animals sticking out. One even managed to lose two corpses without realising it. What worried them most was that farms either side of the A470 started going down with the disease until by June there was no living stock in the entire valley. When, weeks later, there was another isolated outbreak, it is hardly surprising that many critics were deeply suspicious that MAFF itself, with its pyres, pits and wagons, was causing the spread of the disease.

There was little enthusiasm for mass burial in Anglesey either when the local population learned that Penhesgyn landfill tip was designated to take up to 10,000 carcasses. People packed out a public meeting to show their opposition to the plan, not least because the Environment Agency had had to launch an investigation after 'body fluids' from mass graves on the island leaked, turning nearby streams red.

The people of Devon watched what was happening elsewhere in a state of acute apprehension. They were, to use the popular phrase, between a rock and a hard place, suffering by far the worst backlog of dead animals but fearful of what they had seen and heard of what was going on at Longtown and the Epynt. It was following the Easter bloodbath that they learned of the plan to build ten massive pyres close to the town of Holsworthy in North Devon, and of the intention to excavate an enormous burial pit at Ash Moor, near Meeth in West Devon.

The proposed pyres were so close to Holsworthy town that, had the wind been in the wrong direction, it would have been necessary to evacuate both the school and cottage hospital. Protests began almost at once. Yes, urgent action was needed to remove the slaughtered animals but surely it was possible to find a site that was not actually immediately adjacent to a town? The reply from MAFF was that the land had actually been leased to them by a local farmer, who must know what he was doing.

As was the case elsewhere, local people mobilised themselves and well-publicised protests and public meetings began. One very real fear, already raised many times before and so far without an official answer, is whether or not the smoke from the burns might actually be spreading the virus. A Devon correspondent to the *Telegraph*, writing on 16 April, voiced his own fears based on the experience of a neighbour who found himself living close to a smaller local pyre.

> He sequestered himself and his family in their farm at the outset of the current foot-and-mouth epidemic. He shut his pedigree herd of Charolais into their barn and put strict disinfection procedures in place. On the Saturday before last, MAFF ignited a pyre of infected carcasses at a nearby farm upwind of his farm. He and his wife came out into his yard in the early evening to find a thick pall of smoke with blackened particles, some of considerable size, obscuring the view across his farmyard and smothering the building holding the Charolais.
>
> The stench was revolting and the smoke remained for several hours before the wind changed. Forty-eight hours later, one of the Charolais showed symptoms. They had finished lambing their ewes 24 hours before. All the animals are now dead and my friend has been told by MAFF that if he buries the carcasses in a site on the farm he had already identified, he would be prosecuted. MAFF will start another pyre, using his stock tomorrow . . . You do not need to have extraordinary intelligence to realise that if you set infected carcasses to burn, at least until the heat is built up there is every likelihood of the virus being carried in the smoke. If Blair or Murdoch or any other jobsworth dares to repeat the smear that the farmers are responsible for spreading the disease down here, his prospects are likely to be as short as those of the unfortunate animals.

Over the next two weeks there were threats of legal action, more marches, more meetings but in the event the pyres still went ahead. Alan Marshall, the farmer who had been coordinating the action against the

Holsworthy pyre, rang the office of the Chief Medical Officer before the burn started, pointing out that the Longtown pyre had been stopped due to health risks. 'He said it was a different type of burn. But the only thing different was the timescale. He could not tell me what lesser pollution problems there would be and his total grasp of carcass burning seemed somewhat limited. Our court injunction was then granted with legal aid, but withdrawn at the last minute, which leaves a foul smell in already overburdened nostrils. I am now of the opinion that Big Brother has well and truly sat on the people of Holsworthy.' Local television viewers were also treated to various officials assuring the public that the mass burning of animals was no more different or detrimental to public health than what took place annually on Bonfire Night.

The Holsworthy protestors achieved a limited victory, only five of the ten pyres were actually used in what was described as a 'limited burn', although the images of the burning, twisting corpses were very dramatic set against the night sky. Also, during the first week only 3,800 carcasses were actually burned compared to the 10,000 proposed. The site foreman had been instructed to go ahead only if the wind was from the south-west but the pyres were actually lit when the wind was blowing from the north-west and the smoke only just missed the town.

It was not until May that MAFF unveiled its plans to excavate the biggest ever burial pit on Ash Moor just outside Hatherleigh, designed to hold up to half a million carcasses possibly from all over the country. This time the protest started at once with some protestors blockading the road to the site and a substantial number mounting a 24-hour vigil in shifts. They felt, said one, that they were being made the subject of 'a massive and very dangerous experiment'. Environmentalists complained that the site, an old clay pit, was unsound, consisting as it did of permeable clay and shifting sands; there were also the remains of old mineshafts in the area. The proposal was to pile in the rotting carcasses, wrapped in plastic and stopped in clay, and subject them to a process known as anaerobic liquefaction.

This appears to mean that the bodies would slowly turn into what one protestor described as 'a stew of putrescence called leachate, complete with toxins such as E. coli 157, ammonia, faecal streptococci,

BSE and many other horrors'. MAFF claimed that the leachate could be contained with plastic liners for 4–24 years. Environmentalists pointed out that as the bodies decomposed and the leachate was drawn off, the clay on top would subside imposing huge pressure on the plastics. Liners used in a similar site in Worcestershire were already leaking. And what about airborne pathogens? Nobody seemed to be able to answer that. Or the gasses MAFF were proposing to flare off? And the 17 lorries a week drawing off the 25,000 litres of leachate from the 15 sections of the pit? How many toxins and what vile smells would the local inhabitants be exposed to?

'Beef is perfectly safe to eat?' protest leader Ron Dawson told the *Western Morning News*. 'Who said that? MAFF did. And now, that same inept organisation is proposing to conduct a huge experiment on our doorsteps.'

At the time of writing the Ash Moor pit still remains unused. It appeared from what Nick Brown had to say, before he was sacked from office immediately after the general election, that it might never now be needed, but the new Minister, Margaret Beckett, who took over at the new Ministry, DEFRA in June, has refused to rule it out should the worst happen and the disease flare up again in the autumn. MAFF itself revealed in official figures on 22 May that Ash Moor has, so far, cost the taxpayers £7.5 million.

# Chapter 7

# Taking on MAFF
# – The Losers

Blair was now working towards a general election on 7 June, and the official announcement of the date was to be made during the first week in May. Nothing was to be allowed to stand in the way. Towards this end a spate of positive publicity poured out of Millbank: the major efforts being made by everybody concerned; the splendid role the army was playing; the financial resources being poured into the campaign to beat the disease once and for all; and the fact that, yes, the number of cases per day was coming down. In Devon, thanks to the assistance of the army and an increase in the number of rendering plants taking carcasses, the backlog of rotting animals was at last being reduced. Very little on the subject was now being heard from either Scudamore or Brown (except when he was wheeled on to repeat his usual mantra). Chief Scientist David King was now the most prominent spokesman for government policy, brought out at every opportunity to explain it and defend the contiguous cull. No one in the media questioned his credentials. He was the expert, the Chief Scientist.

However, just as there is criticism over the role of the Imperial College team under Professor Roy Anderson, whose advice led to the policy of contiguous cull (criticism which is specifically directed at their lack of veterinary knowledge and experience of animal husbandry), attention is now being drawn to the fact that King's experience in this particular field was also strictly limited. He is a chemist and his expertise lies in his work on surface metals. As a major livestock auctioneer put it: 'Shot through this terrible time has been the blind faith Blair has put in Professor King ... a chemist with a speciality in surface metals. So on that basis, if I have trouble with rust on my car, I should take it to a vet!'

But it would appear from Channel 4's *Dispatches* programme, transmitted on 4 July, that Blair, infuriated by MAFF, Scudamore and Brown, was now only prepared to listen to King, the Imperial College team and NFU leader Ben Gill. Of the latter, one disenchanted Cumbrian farmer commented that he presumed Gill was 'waiting for the accolade to fall on his shoulders, which is not what Cumbrian farmers have in mind to fall on his shoulders'.

The *Dispatches* report was an extraordinary programme, however, in most respects. After seeing it, the average viewer could be forgiven for thinking that only Cumbria had really suffered as a result of the outbreak, since no other part of the country was mentioned; that the only people that mattered with regard to policy and action had been King, Gill and the Imperial College team; that everybody who had been affected – farmers, country dwellers, the tourist industry – had been, presumably, content with what this team had achieved since, apart from one farmer, no one else was asked to give their views. There was barely a hint of the major controversy raging throughout the countryside. All in all it was a great missed opportunity, many – if not most – of those closely affected were very disappointed at what was trailed as 'the truth' behind the epidemic. Instead, what we saw was the metropolitan view of the disaster that has dogged it from the start.

Towards the end of April, the question of vaccination was raised again and it was even being suggested that Blair was coming round to the idea but was unable to go ahead unless 'the farmers', that is the hierarchy of the NFU, were kept on board. Other sources say Blair did not favour vaccination and was annoyed that it should ever be thought that he did. An edition of *Newsnight* at about this time opened with a shot of Blair and Gill together on the steps of 10 Downing Street.

Jeremy Paxman's voice-over informed viewers that one of these two men had the power to initiate a programme of vaccination and possibly stop the epidemic in its tracks. 'The other is the Prime Minister.'

Towards the end of April and beginning of May a curious thing happened. It is reminiscent of the famous exchange between Sherlock Holmes and Dr Watson when Holmes muses on the strange case of the barking dog. But it didn't bark, returns Watson. That, Holmes informs him, is what is strange. Suddenly, after weeks during which the epidemic had led the news agenda, like a tap which has been turned off, the foot-and-mouth epidemic virtually disappeared – with the honourable exception of *Channel 4 News* – from the national media, though it must be emphasised that this was not the case where regional newspapers were concerned.

Where had all the headlines gone? Gone to clear the way for the general election, every one. For, as far as the epidemic was concerned, Blair told the nation, we were now on the 'home straight'. All that was required now was the starting pistol. On 7 May he finally delivered the long-awaited election announcement and went to Buckingham Palace to request permission from the Queen to dissolve parliament. The new date for the general election and county elections was to be 7 June. On 8 May, MAFF promptly cancelled all press conferences on the subject of foot-and-mouth for the duration of the election campaign 'in accordance with Cabinet Office guidance'. It must have been a relief for them as they were already under considerable fire for removing essential information about the numbers of animals caught up in the contiguous cull and their disposal from their website. It was claimed this was done because MAFF were concerned the figures might not be sufficiently accurate . . .

The local media were up in arms. In the south-west the editors of the four daily regional newspapers and the local evening and weekly papers joined forces with local radio and television news editors to demand the reinstatement of MAFF's figures. In an open letter to Tony Blair, they emphasised that they were fully aware that during election periods civil servants must remain completely neutral and not make statements which might be considered to reflect credit on any one particular party, but they found it impossible to understand how this could possibly be applied to the present emergency situation.

In the West Country, they pointed out, culling was still proceeding amidst growing disquiet among many people about the wisdom of the policy; there were still animal welfare issues which were of deep concern; and there was still confusion, uncertainty and unrest regarding the disposal of carcasses, especially the excavation of the vast burial site at Ash Moor. In the foot-and-mouth-affected areas, it continued, there remained a pressing need, in the public interest, for questions to be asked on policy and the public relied heavily on the regional media to act on their behalf as a watchdog.

'Because of the decision to suspend press conferences and provide only "factual information" we cannot fulfil that role until the election is over. That is, whether by accident or design, an intolerable obstruction to the public's right to know and we cannot believe that, given your frequently stated commitment to such rights, the situation which we are now in is what you really want. Please restore full media freedom, rights and access as a matter of urgency so that we can continue to serve these people.'

The response from the MAFF's deputy director of communications, Lucian Hudson, to the editor of the *Western Morning News* on 10 May was that their 'outrage' was misplaced. 'Most journalists understand and accept the long-standing restrictions placed on civil servants during an election campaign.' There was not, as had been inferred, any cover-up or news blackout. Questions on technical and operational matters (but not policy) would be answered and, where appropriate, MAFF would contact farmers directly on important disease control issues 'as we have done throughout this outbreak'.

It was against this unsatisfactory background that those individuals prepared to fight the policy of contiguous cull had to make their stand and, while the regional press gave them every support, it was only a lucky few who were to make it into the national media. Among these, before the national media curtain fell, had been Phoenix the Calf. Phoenix was the little white calf found alive after a whole pedigree herd had been slaughtered and the fight was on to preserve her from the cull. There was widespread national publicity as her owners hung on in there, beating off all slaughter attempts until the magical 21 days had passed and on 24 April she was pronounced clear of infection. It was claimed that Tony Blair had intervened personally and that because of Phoenix the government 'relaxed' its contiguous cull policy. 'Relaxed' was, however, a somewhat strong adjective to use. It was, claims Anthony Gibson, merely 'tweaked'. If the policy had indeed been relaxed then there was precious little sign of it on the ground. But the headline-writers loved it and it was good publicity for the government at the right time.

The election and the dropping of the MAFF's press conferences was extremely convenient for the government because it meant that the growing opposition to the policy of contiguous cull was effectively silenced and the more dubious activities of MAFF employees could not be publicly questioned. For by this time, from every single district which had suffered from the epidemic, the same story was heard from farmers time after time and far too frequently to be dismissed as mere anecdotal evidence or hearsay. The story was that those farmers who tried to resist the cull were being threatened with a variety of sanctions if they did not agree to it.

'Farmers were ringing in here,' says Anthony Gibson, 'saying "I'm being told that if I don't allow the slaughter teams in, then I won't be paid compensation."' That was the most common threat and it was peddled in Cumbria, Dumfries and Galloway, Wales, Hereford and Worcester, Warwickshire, Yorkshire and Cornwall as well as Devon. Hundreds of farmers are prepared to vouch for it, saying that the threat was made so universally that it seemed to be part of MAFF's policy.

MAFF denied there was any such instruction given to their vets and slaughter teams. Whether there was or not, an illuminating leaked document, Annex 1 to FIN 2001/170 08 Briefing on High Court Judgements on Contiguous Culling Policy, throws at least some light on the instructions that were definitely given to MAFF vets and slaughter teams. The purpose of the document, written by the Deputy Director of Foot-and-Mouth Operations at MAFF Headquarters and circulated to Regional Operation Directors (RODs) is: 'To provide RODs and other officials with information to enable them to seek to discourage farmers from resisting MAFF's slaughter policy by citing the favourable judgements of the High Court. The strong possibility that farmers will be liable for MAFF costs in future cases should also be used as a deterrent.' It then spells out the arguments:

> Judgements of the High Court on MAFF contiguous culling policy. Key points:
>
> The High Court has decided that the Minister has the power to slaughter an animal he thinks has in any way been exposed to FMD.
>
> The Minister's powers are not constrained by EU law, which sets out the minimum reaction required of a member state.
>
> One farmer has already withdrawn his case in the light of previous judgements.
>
> It is highly likely that in future cases, MAFF will be awarded costs if successful.
>
> MAFF has successfully obtained injunctions backing our proposals to cull animals on contiguous premises in two cases (Winslade and Jordan). In one further case (Baines) the farmer dropped his opposition to slaughter of his animals after MAFF Legal Department wrote to his solicitors explaining that MAFF has legal powers under domestic and community law to cull his animals and that this has been accepted by the High Court in two recent cases. High Court Judgements have confirmed that under Animal Health Act legislation, the Minister has the power to slaughter if it appears to him that 'animals may have been in any way exposed to the infection of foot-and-mouth disease'.
>
> It is not for the Court to substitute its own view for that of the Minister, merely to judge whether or not his view is reasonable and whether or not he has exercised his powers lawfully under the statutory provisions.

There is much more in the same vein emphasising that the cull policy is legally permissible under both national and European legislation. 'It is not for the Court to second guess veterinary judgements on whether animals are susceptible to the disease.' The writer of the document has to admit that in the two cases cited above, Winslade and Jordan, the court did not award costs to MAFF, but concludes, hopefully that 'it is very likely that the court will now award costs in future successful cases'. Farmers take note.

One recipient of a copy of the document was so stunned that they took legal advice (being in the fortunate position of easily being able to do so) and showed it to both a solicitor and a barrister. The solicitor's reaction 'was one of shock and horror'. Both lawyers consider the information given in it to be seriously flawed. Although quoting two court judgements (Winslade and Jordan) and a third case (Baines) the document fails to emphasise that each case was judged on its own merits. 'A very important point.' We all have a right to have our particular case heard and judged in a fair trial. In the first two cases, no alternative expert advice was offered to the judge; this is important as, without such evidence, there was nothing he could do but find in the Ministry's favour. It is interesting to note that when 'expert evidence' has been offered, MAFF has withdrawn their action and so no judgement was obtained.

There also appears to be another serious error in the document, namely a misquoting of the legislation. The Animal Health Act 1981, Schedule 3, Paragraph 3 actually says: 'The Minister may, if he thinks fit, cause to be slaughtered any animals which appear to the Minister to have been in any way exposed to the infection of foot-and-mouth disease.'

You will note, goes the legal advice, the replacement of the word 'may' by 'to'. There is surely quite a difference between animals that one suspects 'may have been exposed' to ones that 'appear to have been exposed'. The Act is far more specific than the document. 'It is one thing for MAFF/DEFRA to provide its officers with information, it is another thing to suggest that they seek to deter affected persons, who may be in a very vulnerable position, from exercising their right of access to the Court.'

The mass slaughter policy was to affect those involved in many different ways. It may well be that there is some truth in the view that for the bigger agribusiness farmers it meant very little and that at every bang from a stun gun, they calculated just how much more they were going to receive in compensation. For some of those involved, such as Willie Cleave who received around £700,000, the answer was a very great deal. For others the situation was almost more than they could bear.

The antithesis of the compensation seeker was a farmer in Roseburndale Valley in Cumbria who lost his livestock due to one nearby case of infection. He left his own note on a website. 'Prior to the slaughter I also gave the sheep some Bach Flower remedy called Aspen which is for fear of death . . . The remedy was sprayed on the sheep so it touched their eyes and noses. Within minutes of using it, the sheep and lambs settled peacefully in their pen. I took this remedy as well and so did the other participants. I hope that this will help other farmers cope with this devastating time.' As the slaughter began he blessed his flock with the Celtic blessing:

> May the road rise up to meet you,
> May the wind be always at your back.
> May the sun shine warm on your face.
> May the rain fall softly upon your fields.
> And until we meet again,
> May God hold you in the hollow of his hand.

As the story above would suggest, people who find themselves in the desperate position of trying to save their animals are in an extremely vulnerable state. The stories of some of the winners and losers in the battle demonstrate this only too clearly. As the Winslade case was one of those claimed as a success by the Deputy Director of Operations in the MAFF document, it might be as well to start with a few representative examples of those who lost the battle against slaughter, and to show just what was authorised in the name of the Animal Health Act 1981, brandished by MAFF employees at the farm gate.

### LES AND GRETA WINSLADE – THE £30,000 THREAT
Les and Greta Winslade are organic farmers living near Knowstone in Devon. Their problems began when Les Winslade noticed that a flock

of sheep were running free and were now in a field belonging to a neighbour. He promptly informed MAFF of what was happening. There was no response. For days they did nothing, even though he rang them again, twice in one day. When the response did finally come it was to inform the Winslades that MAFF were seeking an injunction to kill his dairy herd under the contiguous cull policy.

The Winslades had a herd of 60 healthy organic Charolais suckler cows and their calves and they were determined to make a stand, not least because their information had been ignored for so long. To that end, therefore, during the third week in May, by which time the general election campaign was in full flood, they barricaded themselves into their farm and prepared to repel MAFF vets, slaughter teams and, if necessary, the police. They rightly thought they had a fair chance of winning the battle since, in another highly publicised case, that of the Thomas-Everards, MAFF had finally backed off when it became obvious that the animals were provably uninfected after the maximum incubation period of 21 days.

'If I were a hard-headed businessman,' said Les Winslade, 'I'd probably be letting the slaughterers go ahead and taking the compensation. But I'm not. I'm a farmer. I had two sets of twin calves this year and I struggled to keep them alive, and I look at them and I think, why should I let these people kill them when they haven't even had a life?' MAFF vets had been regularly monitoring his cattle during which time they showed no sign of infection and the nearest infected farm was over half a mile away. Time passed. Les padlocked his gates, refused access to MAFF valuers and parked his tractor across the entrance. Protestors gathered outside to support his action, including the successful victor of the Thomas–Everard siege.

The Winslades were worried from the start that they simply did not have the financial clout to take on MAFF in the courts. As Les Winslade told the *Western Morning News* on 23 May, his income had been so poor the previous year that he had not even earned enough to pay taxes and MAFF had told him that if he went to court, he would have to pay all the costs, but eventually an appeal was lodged. By now there were only four days to go before the herd reached the end of the 14-day 'minimum incubation' period without showing any sign of infection and might reasonably have then been given a further seven days' grace.

The siege was taking its toll on Greta Winslade who lost half a stone in just over a week as she was unable to eat for anxiety. 'I refer to MAFF as Hitler,' she said. 'And the thought of them culling the baby calves is like being invaded by the Germans.' Leila Winslade, the couple's daughter, described her father as 'an honourable and proud man', who was likely to face financial ruin as a result of the stand he was making. 'This is political now,' she said, 'it's not just about my parents' animals. If MAFF lost this case they would face millions of pounds in compensation to farmers whose livestock was unlawfully culled – it's much bigger than my poor old dad, though he may be ruined by the legal costs and he is petrified about them.'

To add insult to injury, Les Winslade had been astounded to be rung up by a MAFF official to arrange a time for 'the disinfecting team' to come in to clean his farm, following the cull of his animals on 15 May. 'But I told him nothing had happened on 15 May. It is unbelievable. When it comes to MAFF, the right hand doesn't know what the left is doing.' A spokesman for the solicitors the Winslades had employed to appeal on their behalf confirmed that MAFF had not yet even issued the relevant court injunction and a MAFF spokesman then apologised for their 'administration error'.

Then on 23 May, without warning and before the result of the appeal was known, the slaughter team arrived, backed up by police in case of protests. The stand-off lasted throughout the day as protestors gathered in support of the Winslades and their herd of organic Charolais cattle, many of them pregnant cows. But eventually the Winslades gave in, unable to bear day after day of similar stand-offs if MAFF still refused to accept that the animals were not infected and culled them anyway. Among the supporters was John Gouriet of the pressure group Freedom in Action who had been taking a major role in supporting farmers who wanted to take a stand against slaughter. His reasonable suggestion to MAFF was that they keep the animals under observation until the end of the incubation period then give them a final test. But at five o'clock that afternoon he came to the farm gate and told the protestors that the Winslades could not fight any longer. 'Our lawyer's advice is that there is no realistic possibility of persuading

MAFF not to go ahead with the cull.' Gouriet believed they had a strong case and that MAFF's response, that this was the only policy and it had to be carried out, was provably untrue as there had been substantial variations in different parts of the country and in other cases. The pressure on the Winslades had been 'merciless', however. Gouriet described it as 'a terrible day for freedom and democracy in this country'.

Protestors were appalled. 'They have taken advantage of gentle folk by doing this,' said Jess Barbery of STAMP – Stop the Ash Moor Pit. 'I will be 60 soon and never thought I'd see anything like this in my lifetime. There are newborn calves in the fields which will soon be dead.' Another protestor turned to a policeman saying, 'You have no morals and you have no courage, if you did you would protect all the citizens in this country.' The police responded that they were there to ensure people could go about their lawful duty, including protesting.

The Winslades had also been given a brutal ultimatum by MAFF. Let them in to slaughter the herd or face legal bills of over £30,000. Major landowner Hector Christie, of nearby Tapeley Park, speaking to the *Western Morning News* on 1 June, described the way the Winslades were treated as 'disgraceful' and the threat as 'truly abominable. I was there as MAFF officials barged their way on to the farm and it was the most alarming and depressing sequence of events I have ever witnessed. Since then I have been determined to ensure the government will be accountable for their actions during this crisis.'

GORDON AND THELMA WILLMETT – COMPOUNDING
A BUNGLED CULL

Knowstone was also to become notorious in the annals of the epidemic as one of the places where a cull was carried out in the most appalling manner – a cull that was to lead on to more contiguous culls. On 13 May, during the slaughter of livestock on Barton Farm, 20 Limousin bullocks broke loose and ran amok across fields on neighbouring farms. The result was that local television viewers were treated to amateur video footage of gunmen taking pot-shots at panic-stricken cattle. Sixteen were killed on a nearby farm but four remained alive until the following afternoon when the army were called in. Two were then killed on a nearby farm, the third was discovered and shot on yet another, and a fourth after being tracked down to a copse.

Asked to comment on it Anthony Gibson said: 'MAFF bundled into the herd of Limousin bullocks with men in white suits. Anyone who knows anything about farming will know that Limousins are flighty at the best of times and, surprise, surprise, the animals bolted in all directions. They then pursued them with rifles.' He continued, 'The whole thing was mismanaged from start to finish . . . You would have thought that after having over 160 cases in Devon alone, they would have learned how to deal with the situation.'

Gordon and Thelma Willmett had already lost nearly 100 sheep because they might have come into contact with the bullocks while they ran through one of their fields, but their cattle had been safely locked up out of harm's way and could not possibly have come anywhere near the bullocks. Thelma Willmett, therefore, was determined to save her own eight cattle and the 25 belonging to her daughter-in-law, insisting that there was no point in culling them so near to the end of the incubation period and they were showing no signs of the virus.

Protestors gathered outside the Willmetts' farm, using vehicles and a boat to block the lane leading up to it, and on Sunday, 3 June, managed to prevent MAFF slaughter teams getting through the gate and on to the land. But the Willmetts were aware that time was running out, there could be no indefinite protest. On that first occasion police inspector Geoff Woodley, who had been in charge of MAFF's attempt to get on to the Willmetts' farm, had told the protestors that it was not up to him to judge the rights and wrongs of MAFF policy but that he was prepared to consult with the MAFF team. He then returned to tell the protestors that the officials had withdrawn – for the time being – and the protestors celebrated what they saw as a victory. But it was to be short-lived. At the crack of dawn on Monday morning the slaughter team returned, forced their way past Thelma Willmett and her 71-year-old husband to slaughter the cattle, none of which were infected.

CAROLYN HOFFE – ONLY FIVE SHEEP

In most of the cases that did receive publicity, the lone protestors were farmers with sizeable numbers of animals. Carolyn Hoffe is not a farmer. She lives near the Isle of Whithorn in Dumfries and Galloway in a house which is divided into two (her parents living in one half) and with a paddock in which she kept five sheep. She is recently widowed. Towards the end of April, animals on a farm a couple of miles away were confirmed as having foot-and-mouth and she was warned that her handful of sheep might well come within the contiguous cull area. However, as the days passed and her sheep remained healthy, she began to feel more optimistic. Then, 23 days after the initial outbreak, she learned that there had been another, this time only half a mile away, and that, therefore, her sheep must be culled.

The owner of the farm which had been affected owned a considerable amount of land, some of which was indeed contiguous to Carolyn Hoffe's paddock, but the farmer's animals were a considerable distance away and all but one of his nearby fields contained only crops. She was given leave to appeal against the decision, which she did. Carolyn also caused a stir and secured considerable publicity by promptly bringing her five sheep into her house and putting them in her living-room where they were to live for the next five days, during which time she hoped she would be given leave to seek a Judicial Review. She then ensured that the interconnecting door between her house and that of her parents was securely locked at all times and that the only way the property could be entered was through her parents' front door.

This then was the position when on 4 May, MAFF vets, valuers, police and army personnel arrived in force to tackle one woman and five sheep. At this stage one of Carolyn's friends, Pamela Groves, was allowed into the house to give her some support. Ms Groves had been miked by a Channel 5 news team who had also miked Carolyn's father, Douglas Allen. The first person to demand to be let in was a single female figure in a white suit, apparently by herself, and Carolyn persuaded her mother to do as she asked. As her mother opened the door, a large policeman allegedly forced his way in, pushing everyone aside, followed by MAFF's chief vet in Dumfries and Galloway, Vikki Cleghorn, whose name has become synonymous with culls in southern Scotland. Ms Cleghorn is alleged to have then grabbed Mrs Allen roughly by the arm as she tried to close the door on her. Douglas Allen asked Chief Inspector McCloud, who was in charge of the operation, if he had a warrant and he said that he had, although he would not show it to Mr Allen. The rest of the party then trooped into the Allens' house, one policeman throwing open the door to their bedroom assuming it was the connecting door to the other property. What happened next is a matter of record.

'This is my home,' Mr Allen complained. 'This is private property. You're in the wrong house.' The inspector replied that this was not the case as Mrs Hoffe owned both premises and therefore he had a right to go anywhere he wished. He then demanded the key to the other house and when Mr Allen refused to hand it over, threatened him with arrest. Carolyn and the Allens were then informed that in culling her sheep the officials were acting in accordance with the Animal Health Act of 1981.

The police inspector next demanded that the connecting door be unlocked immediately, saying, 'If you don't, I'll kick it in and it looks quite new to me. Afterwards it won't be a pretty sight.' By this time Mrs Allen was in such a state that Carolyn felt she had no recourse but to open the door, which she did. 'Though I refused to allow Vikki Cleghorn in.' Her father and friend then accompanied Carolyn to the room where the sheep were waiting and they were slaughtered there in the sitting-room in front of them.

At the time of writing Carolyn Hoffe is considering legal action against both the police and MAFF. In an interview with the author in early July she said, 'I look out at the countryside round here and there's nothing left. It's completely empty. I don't know if I'll ever have animals again. Some farms are soon going to be allowed to have sentinel animals – animals brought in to test if the virus is still active round here – but no one has asked me to have one. Anyway I don't know if I could ever have more sheep. What is there to stop all this happening again?'

# Chapter 8

# Taking on MAFF
# – The Winners

The sheer scale of the contiguous cull in Devon, coupled with the kind of examples reported in the previous chapter, provoked an enormous backlash. People who had never taken part in protests in their lives turned out in force to barricade farm gates, block roads and march through the streets. Among the protestors who had a great deal to say was television celebrity Noel Edmunds, who farms in Somerset, and with his support the *Western Morning News* mounted a major campaign backed by a raft of veterinary and scientific experts. This was the Heart of Devon campaign, later to become Heart of Britain, in which Noel Edmunds was to play a substantial role. Day after day two full pages of the paper were devoted to the epidemic in general and the campaign in particular. A website was set up, also a general helpline, a specialist legal advice helpline and an e-mail address to enable readers to communicate with the paper. The *News* also offered readers a free poster showing an elderly farmer with a young girl holding a lamb, bearing the legend 'Stop! YOU Are Destroying Our Lives!'. The response was immediate and overwhelming, enabling the paper to devote whole supplements to readers' letters.

Overall, when it came to fighting MAFF over the cull there were more losers than winners but those who succeeded did so in part because they were able to break through the news blackout on the subject and actually get themselves into national papers and on to national television.

CHRISTOPHER AND GUY THOMAS-EVERARD – LANDOWNERS
SINCE THE NORMAN CONQUEST

The Thomas-Everards (or plain Everards as they appeared in the media) farm beef cattle on the edge of Exmoor near the town of Dulverton: the farm is owned jointly by Christopher Everard and his son, Guy. A couple of weeks before the blow fell, they had been visited by a contract worker, Robert Norman. Norman, who has a farm of his own, also takes on a variety of contract work for other farmers, ranging from acting as a relief milker to dehorning cattle. Norman had visited the Everards' farm to dehorn, or poll, some calves a couple of weeks earlier. He had taken thorough disinfecting precautions and after his visit the calves had been kept separate from the rest of the cattle. Shortly afterwards, Norman discovered his own animals were infected with the virus and the Everards were duly served with a Form A notice confining them to their farm premises, which they were content to abide by. Their cattle were also officially inspected on a daily basis and pronounced healthy. Time passed without any sign of disease. Then, on 4 May, the inspections suddenly stopped without explanation. The first the Everards knew that slaughter of their 980-strong herd was imminent was from friends and the press. No official from the Ministry had called to tell them what was about to happen.

The land has been in the Everards' family since their ancestor, Udo, arrived in Britain with William the Conqueror. Generations of Everards appear on war memorials having died in battle. They were determined to make a stand. They were also certain sufficient time had elapsed, since Norman's visit, to prove their animals were in the clear. They were also enraged that the decision had been made in London without any consultation with local vets.

On 6 May, the date designated for the cull, the Everards prepared to mount a physical defence and a very large cattle lorry was parked across the farm entrance. Guy Everard is a good-looking young man who was about to be married, indeed he had been given special permission to leave the farm for his wedding in Cambridge. He is also extremely articulate and in no time at all the media were gathering at the farm gate to interview him and his fiancée. The Everards also mounted a legal challenge to the cull. Messages of support poured in from across the country, including one from the Rt Reverend Jim Thompson, Bishop of Bath and Wells, who has also had much to say in public about the morality of what was going on. Someone in Dulverton offered to sound the town's old air-raid warning siren when MAFF arrived in order to summon assistance from the inhabitants.

MAFF did not enjoy the publicity. The Everards' quick legal challenge meant that they had to seek an injunction to enable them to go on to the farm as threats of lack of compensation or massive legal fees failed to have any effect in this case. Also, as the Everards' solicitor was prepared to fight to the death, this could take some time. Asked what they proposed to do about this a spokeswoman explained, 'Until such time as we are able to go forward, then there's no action we can take at the moment.' But it was necessary for MAFF 'to see this thing through' and see it through they would.

Christopher Everard's response was to repeat his warning that if MAFF tried to get on to the farm they would 'be surprised at our ingenuity and inventiveness' in preventing this from happening. He pointed out that at the start of the crisis there were three farms on Exmoor in an identical situation yet they did not have to have their animals culled. 'There seems to have been a sudden change in policy and the reaction is blind panic instead of reason.'

By this time the election campaign was up and running and the Everards made full use of this fact. They gave press conferences at the farm gate, musing on why the decision to cull their animals had not been announced locally but at a full MAFF press conference in London before they had even been told themselves. 'It is generally considered proper for a judge to inform a condemned man of his sentence in person,' Mr Everard told the press. 'Contract killers, of course, do not tell their victims in advance and that, I fear, is what the slaughterers will be. It is now 11 days since the supposedly dangerous contact and signs of the disease can usually be detected within three to eight. We are not whingers. We will accept justice and common sense, but this is neither, and we are going to fight.'

Shortly afterwards, with no real explanation to their change of heart, MAFF threw in the towel and allowed the Everards a stay of execution until sufficient time had elapsed during which any infection would have showed itself. On the surface, there seems little difference between this case and those of the Winslades and Willmetts. Except that the Everards were exceptionally articulate, had been major landowners in the area for centuries and had sufficient funds to fight MAFF to the last ditch and were quite prepared to do so.

STEVE PHILLIPS – DOING IT FOR LINDY

On the other hand, Steve Phillips did not have much in the way of financial resources, or ready access to the media and the ability to use it to his advantage. He was one of the unfortunate farmers caught up in the cull following the bungled slaughter at Knowstone. He had only seven sheep on his farm, all of which had been lovingly reared by his partner Lindy who died of breast cancer five days before he barricaded himself into his farm. 'I'm doing this for me and Lindy. I'm sure she is with me in this. We are fighting for everybody who has been screwed up and had their lives ruined by MAFF,' he told the *Western Morning News*. Because he had barricaded himself in and dared not leave the premises, he was unable even to arrange a funeral for Lindy whose body still lay in the local chapel of rest.

With an effigy of 'a faceless MAFF man at his gate', Phillips sat in an armchair at his farm entrance and refused to let anybody past. 'Somebody has got to make a stand. The way this village has been pushed round and bullied by MAFF is not right. I will be here tomorrow, the next day and the day after that – and so will my sheep. I am placed here in an untenable situation. I have got other livestock that need tending to. I have cattle calving and I can't tend them properly.'

His plight aroused great public sympathy and the response of MAFF did them little favour in the field of public relations. Asked about this specific case a spokeswoman (possibly the same as had answered the query regarding the Thomas-Everards) intoned: 'All livestock has to be culled in line with the national disease control strategy.'

'I won't allow them in if they can't prove that the animals are positive,' responded Phillips. 'If the blood tests are positive then I have got no problem with them coming in, but I've got to say that you can't trust anything MAFF say. They lie to everyone.' Three days later MAFF backed off, the news being brought to Steve Phillips by Christopher Everard, who had been acting as an unofficial adviser to others following his own victory. Phillips' reaction was one of cautious relief, cautious because MAFF was 'so slippery'.

MOSSBURN ANIMAL CENTRE – SAVED AT THE ELEVENTH HOUR
Juanita Wilson has been running the Mossburn Animal Centre at Hightae, near Lockerbie, for 13 years. She and her staff take in, and look after, unwanted cattle, pigs, sheep and goats. Towards the end of March there had been several cases of foot-and-mouth in the area, after which there was nothing. MAFF action, she says, on the nearest farm to the Centre appears to have been exemplary: the animals were culled on the Friday and the carcasses taken away the next day. So far so good.

She has only three sheep and they were in a field adjoining a concrete yard. 'Since I'd never seen foot-and-mouth in sheep, I asked for the sheep to be examined and a vet came, thoroughly examined them, and passed them as fit and healthy. They were later seen by a lady vet on secondment, who again pronounced them fit although she did not examine either their feet or mouths.' But the vet told Juanita Wilson that her biosecurity 'was very good'. She was, however, served with a Form D restriction ('which I expected') and the Centre was closed to visitors.

On 30 April she helped organise a meeting calling for the policy of the contiguous cull to be ended in favour of vaccination. It all went very well and over 200 people packed the local Academy to hear speaker after speaker call for a change of policy. A few days later, at eight o'clock in the evening, she received a telephone call that was 'to initiate 12 days of hell'. It was an official from the Scottish Executive telling her that her animals were to be culled as she came within the three-kilometre contiguous cull zone. She received no answer to her query as to why, if that was the case, the animals had not been culled six weeks earlier at the same time as other farms in the area. 'If the ministry wanted to include my animals in the pre-emptive cull, that was surely the time to do it.' If her animals were going to develop the infection they would definitely have done so by then.

She is now convinced that the action was taken because of her participation in the Lockerbie meeting – 'at which I had made my feeling about the pre-emptive cull very clear: that I am totally opposed to it and that it is disgusting and probably illegal'. It was, she says, 'payback time'. The next morning she contacted her solicitor and asked him to take out an interim interdict so that the case could go to appeal. She wanted it heard as a devolution issue, so that if the appeal was turned down it would still be 'live' in the Scottish courts, but in the event the devolution issue was not raised and she lost. 'It had all been a complete waste of time.' Determined not to give in, she turned to the media and soon support came from all over Britain. A protest group, Against Sanctuary Killing (ASK), was formed and within days over a hundred protestors were camping out on land nearby, accompanied by television and satellite vans.

More or less convinced by now, however, that nothing more could be done, Juanita Wilson did a deal with MAFF. They would come in at ten o'clock next morning to organise the slaughter ('not cull'), but she insisted first of all that she would be with her animals and that they should not die one after the other in front of each other, and secondly that her own vet would put them down. MAFF agreed to this. She then told the protestors there was no point in continuing with the protest and a good many began to drift away. But as time went on 'I began to have a gut feeling that I was wrong'. Then, at seven o'clock in the evening before the animals were to be slaughtered, she had an unexpected call from a legal sympathiser.

'Don't trust MAFF,' he told her, 'come up to Glasgow and I'll arrange for you to swear an affidavit asking the Court of Session to review their original decision not to grant a Judicial Review. With her driveway blocked by her car, to prevent MAFF entering the premises while she was out, she and her friend, John Dowson, drove to Glasgow. After a lengthy discussion with their legal sympathiser, at three o'clock in the morning they knocked up Notary Public Michael McSherry who, in spite of the time, agreed to draft the necessary affidavit. This was to go direct to Edinburgh and Ms Wilson and Dowson then drove back to Mossburn, arriving at 4.30 a.m. armed with photocopies of it. Shortly after they returned, John Dowson set off for the MAFF 'bunker' in Dumfries to deliver a copy there and also to Dumfries and Galloway police headquarters, while Juanita went to bed to try and catch up on some sleep.

But within a short time she was up again, having received a call that people were 'pouring out of the MAFF bunker' and, accompanied by police, were on their way to Mossburn. John Dowson meanwhile was having considerable difficulty serving the affidavits. He had no joy at all at the bunker, his first port of call, then went on to police headquarters. He reached the building just after five o'clock, by which time the MAFF team had almost reached Mossburn. He asked to speak to a senior officer, was kept waiting for some time, but finally managed to serve the affidavit on an Inspector Murray. He then went back again to the bunker and this time roused two officials, one of whom was a vet.

MAFF and the police arrived at Mossburn in force. The word 'overkill' comes to mind. Janet Rogerson, who lives in nearby Heck, was stunned to see a convoy of police vehicles en route for Mossburn. One witness also claims it included a prisoner transport van. 'At Mossburn police cars were parked diagonally across the roads and the main Dalton to Lochmaben road was blocked from 5.50 a.m. to 8.50 a.m. At one stage there was a solid line of police cars on the main Lockerbie to Dumfries road.' 'Numerous local people were prevented from going their normal route to Annan,' claims Ms Rogerson, 'and a care worker in uniform was not allowed through, leaving her client waiting.'

Once at Mossburn, police climbed over Juanita Wilson's car and an inspector informed her that the cull would now take place. Not only that, she would not be allowed to have her own vet put the animals down, this was to be done by a MAFF vet. When she asked about the enormous police presence, an inspector informed her that they were there 'to maintain law and order'. She then handed the inspector a copy of the affidavit and told him 'MAFF will not be coming in here'. He replied that he would have to take it away and seek legal advice.

Meanwhile complaints were pouring in about road closures. John Dowson, who is an ex-leader of Dumfries and Galloway council, protested that there was a road safety issue, as every road around the Centre was now closed. The stand-off lasted two and a half hours, then Juanita Wilson had a call from an Edinburgh solicitor to say that the appeal was going to be heard that afternoon and that she had to be in Edinburgh by two o'clock. Finally the MAFF operation was scaled down and she was able to drive to Edinburgh to the Judicial Review Court, a court in which only Queen's Counsels are allowed to speak. The Petition was accepted by Lord Justice Clark, who ordered a hearing for the following Tuesday. MAFF immediately lodged an objection, demanding the hearing be brought forward to Saturday. When that was refused, they insisted it be heard on Monday. The QC for the Scottish Executive, Gerard Moynihan, opposed any further delay, telling Lord Clark that 'he could not give any undertaking that the cull would not proceed'. Lord Clark replied bluntly that since the culling had already been delayed for six weeks, three more days would hardly make any difference. The matter, he declared, would be heard on Tuesday.

At 9.30 a.m. the following morning Juanita Wilson learned on local radio that Minister for Rural Development Ross Finn had announced a change of policy. There would no longer be an automatic cull within three kilometres of infected farms. She was off the hook. Among the first to congratulate her was Carolyn Hoffe for whom the new ruling had come too late.

Juanita Wilson has been left with a legal bill of £15,000. Also many questions remain, not least those regarding the sheer scale and cost of the operation mounted against the Mossburn Animal Centre. Superintendent George Graham of Dumfries and Galloway police, in a statement made the day after the Minister's decision, said: 'The temporary road closures were in the interests of road safety and have been a feature of operations where culls have taken place. The restrictions were lifted as soon as it became clear that there would be no cull at that time. As with all police planning, we had to be prepared for a whole range of possibilities. We were delighted with the peaceful and dignified way the protestors made their point. As a force, we are proud of the good relationships we have with all our communities no matter what their views.'

Juanita Wilson said she was over the moon at the victory 'but my thoughts go out to all those people who have unnecessarily lost animals because of this evil'. Her fight against MAFF has cost her over £15,000, money she simply has not got but must somehow raise. 'But thinking it over now, I do wonder if it wouldn't have been better, for the sake of other similar cases, if it had not gone all the way and gone to court.'

# Chapter 9

# 'Senseless, Brutal, Inept . . .'

As far as the foot-and-mouth epidemic was concerned the general election was a godsend. The Labour Party leadership was determined to concentrate its campaign on the public services, which meant primarily health and education; even transport, which is on its knees, hardly got a look in. Meanwhile the Conservatives banged on about saving the pound, tax cuts, asylum seekers and law and order. Therefore everything else, as far as the Labour Party were concerned, could wait until after the result of the election by which time, it was fervently hoped, the crisis would be over.

Mark Daniel of the *Western Morning News* put it well:

> The foot-and-mouth epidemic started 'under control' –
> government control, that is. As bloated stinking carcasses
> lay exuding virus in farmyards and fields, as farmers
> clamoured for vaccination firewalls as advised by all the
> experts, as the countryside was smothered by palls of
> foetid smoke, as tourists cancelled bookings and rural
> businesses gazed into the abyss, the epidemic was always,
> officially, 'under control' and the countryside . . . 'open for
> business'.

Then, too late, far, far, too late, MAFF developed a policy which must surely work, a policy which must arrest any plague, just as the spread of any fire must also be arrested by blowing to smithereens anything in the path of the flames. The men from the Ministry justified the brutality, randomness and ugliness of their policy by maintaining that it was working. It was not. The result was wholesale carnage.

He goes on to point out, as have others, that 'culling' is a euphemism. Indeed it is a misuse of language, for hitherto culling has referred to the selective reducing of numbers of animals, not wholesale slaughter, 'not shooting, battering or drowning anything living within a three-mile radius'. *Chamber's Dictionary* definition is: 'Cull: To select, pick out, an unsuitable animal eliminated from a flock or herd.' The relevant section of the law on the slaughter of animals reads: 'No person shall stun, slaughter or kill, or cause or permit to be stunned, slaughtered or killed, any animal without restraining it in an appropriate manner in such a way as to spare it any avoidable pain, suffering, agitation, injury or contusions' – Statutory Instrument No. 731 Animals, Prevention of Cruelty, The Welfare of Animals (Slaughter or Killing) Regulations, 1995.

There is no doubt that much of the work of slaughter, however horrible (or subsequently proved to have been unnecessary), was carried out efficiently and with as much care as possible. The usual method, which most people do not like to think about, is by lethal injection into the hearts of lambs and calves, and the use of bolt guns on adult livestock. First a bolt is shot into the brain to stun the animal, after which a rod is pushed into the hole to kill it. But, unfortunately, this has not always been the case, as witnessed at Gilwern and Knowstone. One account, allegedly written by a soldier serving with the Green Howards in Worcester, which appeared on a variety of websites, suggests that there were instances when matters got totally out of hand. He wrote:

We were briefed that we'd be 'clearing up' – burning

and/or burying carcasses of animals humanely destroyed by trained vets and slaughtermen. But that's all turned out to be more spin and propaganda. What we're actually doing is 'mopping up', killing animals they've left behind or can't be bothered to finish off. My regiment has got all sorts of battle honours for fighting Britain's enemies all over the world, but we're now engaged in heroic hand-to-hand combat with lambs. Their mothers have been shot but some were so frightened by the noise that they escaped all over the place. As we don't have any humane killers, the cleanest way of killing them is just to throw them in the river. We might be trained to kill enemy soldiers, but slitting the throat of a spring lamb or beating its brains out with a blunt instrument, is too much for some of the lads, so they'd rather drown them, even if it's not really as quick.

One of my mates was detailed to stand by a pig which was giving birth. As each piglet was born and crawled away, he had to smash it with the back of a shovel. Once they'd all been born the pig was shot with the others. Worst of all are the cows that have been shot but not finished off by slaughtermen. Some are still crawling around, others clearly still alive but unable to move. We have to beat them to death with lorry spanners or other heavy lumps of iron. If people really knew what was going on, I think there'd be a revolution. The MAFF people say it's the only thing that can be done, but if you ask them why they can't vaccinate, they all come up with different reasons, none of which sound very convincing.

It is impossible to prove whether this is true or not but it is very similar to other stories from other areas, including an incident near Skipton in Yorkshire which was witnessed by at least four people. In this case, it was alleged that a MAFF team in a four-by-four chased cows and calves in it until they were exhausted. They then attempted three times to kill a cow with a bolt gun, without success. Eventually they dragged it up by the neck, hanging it from a small crane device, which finally strangled the cow and killed it. Meanwhile other cows were still running wild and a slaughterman pulled out a rifle to start shooting at them, but put it away when he noticed people watching. The cows were extremely disturbed, making numerous attempts to get to their dead calves. The MAFF team continued to beat and kick the animals while trying to get them into crush crates. According to the witnesses there were about 150 cows and calves. It even proved too much for some of the policemen brought in to support the MAFF officials but they were unable, they told protestors, to intervene.

Incidents like these and everything else to do with the epidemic went on unreported because of the blanket coverage of the election. Those living in urban areas must really have thought that indeed the epidemic was virtually at an end and had been brought under control. Therefore a sudden new outbreak in the area around Settle in the Yorkshire Dales halfway through the election campaign came as something of an unpleasant shock to the authorities, and strenuous efforts were made to play down its severity despite the outbreak rapidly spreading to 17 farms. This time we were not treated to television pictures of heaps of dead animals or smoking pyres although both were present in abundance. However, on 23 May it was suddenly announced that election or no election Nick Brown was to pay a hastily arranged visit to Settle the next day, following criticism from the Bishop of Bradford, the Rt Reverend David Smith, who had accused the government and the media of ignoring the new Yorkshire outbreak.

In a *Guardian* piece on 25 May, entitled 'Minister Brown glum-hands voters in crisis-hit Dales', Simon Hoggart described what happened next. It was, he said, a glorious day in the Dales and if it hadn't been 'for the soldiers striding around, the army vehicles in the car park and the lorries carrying mounds of dead animals away, you'd imagine you were in some Elysium, an English Brigadoon'. The media had been told that Brown would do a walkabout in Settle. 'He did,' reported Hoggart, 'if you count ten yards as a walkabout. He made his way from his car to the town hall door in about five seconds.' As it was he had to push his way through farmers, some jeering, some in tears. 'Others were incoherent, barely capable of putting their feelings into words . . . To distil their arguments, shouted at the lads with notebooks and fluffy mikes while Mr Brown met the NFU, the army, tourism folk and police upstairs, too little has been done too late. They believe that the new outbreak was at first ignored, then under-reported, so that the government could claim foot-and-mouth was nearly finished.'

One farmer told Hoggart that on the day the Labour manifesto came out, and 'Blair was on TV grinning and laughing and smirking', he had said that this was the first day there were no new outbreaks. Yet 12 farms in that one valley alone had been taken out that day. 'That very same day!' The local school had to be shut because for two days no one could teach; the sound of livestock being slaughtered on the hills 'sounded like the assault on Monte Cassino'. Finally the press were ushered into Brown's presence. 'We are bearing down on the disease,' he said about 20 times. He repudiated the charge that civil servants were massaging the figures during the election. 'We are culling out cohorts (neighbouring animals). Nationally we are on the home straight,' he repeatedly claimed.

Another report in the *Financial Times* on the same day mentioned an adviser (who, unsurprisingly, did not wish to give his name), who stated that although the new outbreak was 'serious, it should rapidly be brought under control'. The same spokesman explained that it had all been rather complicated because the disease was not detected for a while and because there had been a large number of animal movements around Settle. This can only be described as a stunning admission. How was this further incompetence allowed to happen after all that had been revealed about the ineptitude of MAFF's response to the outbreak from the start and the crucial need both to recognise the earliest symptoms of the disease and to have the most up-to-date knowledge of the movements of livestock? He admitted that nearly 80,000 animals had already been slaughtered inside the 'Settle Rectangle' alone, but trusted that 'the number of further outbreaks in the area should be low'. This might best be described as the 'crossed fingers' policy of dealing with the disease. One quick wit remarked that 'the rectangle' might as well be called the Bermuda Triangle as far as the government was concerned.

Two days after the Minister's visit, it emerged that an army slaughter team had killed more than a thousand animals 'by mistake', despite protests from the farmer involved. The cattle and sheep were slaughtered at Otterburn Hall Farm, near Settle, after the slaughter team were despatched to the wrong site. The owner of the farm, Philip Metcalfe, had tried to stop the cull but his protests were ignored by the soldiers. He described it afterwards as 'a terrible day. You have no idea what you go through when something like this happens.'

MAFF announced that 468 animals had been incorrectly culled, but Metcalfe said that he had lost over 1,100 sheep and a hundred head of cattle. If anyone doubted his word they could count them. 'They're still lying out there on the ground.' The soldiers had been told initially to kill 280 animals on outlying land used by Metcalfe, which was contiguous with an infected farm. Instead they turned up at his own farm and wrongly insisted that the animals had to die. 'It was a genuine error on MAFF's part,' explained a spokeswoman. 'There was confusion between the army team and a MAFF team,' she said. 'At this stage MAFF can only say that it regrets this action and apologises to the farmer for any distress caused by a genuine mistake.' It is hardly surprising that a few days later a MAFF vet and another official were thrown into the local canal by the enraged populace.

In fact, during the same period of time, there were further outbreaks in Cumbria, the Borders and Devon, which, by rights, should not have been happening if the contiguous cull really was working. Still the numbers rose. By 24 May the number of cases had reached 185 in Scotland and 1,632 overall but no one any longer seemed to be counting. When the media were running with the story the number of animals actually being slaughtered every day had reached 33,000, but by three weeks into the election campaign it had reached 80,000 a day. There have been many suggestions since, which are impossible in the circumstances to prove, that slaughter teams were going in and killing livestock which might well have been infected, but tests were not carried out so that the total of infected premises which was reported could be kept as low as possible.

Still the message came: that the policy based on the Imperial College computer model was the only one that would work. It was as if MAFF and the government, having dug themselves into a hole they could not get out of, saw the only way of escape as digging ever deeper. Also, it was becoming clear that whoever emerged triumphant from the general election campaign was going to be faced with an astronomical bill for compensation. But by this time, even if some of those who had initially promoted the policy were having second thoughts (and it seems some were), it was impossible to admit that they might have been wrong. Too much was riding on it.

In retrospect it does look as if the first doubts began to seep in as early as 18 April, before Blair 'saved' Phoenix. He had received a letter from the foot-and-mouth centre in Dumfries signed by two senior vets, David D. Wardrope BVM&S, MRCVS, TVI, and Roger S. Windsor MBE, MA, BSC, BVM&S, MRCVS, TVI. Copies of it were also sent to the Cabinet Office, Nick Brown, Ross Finnie, two local MPs, Jim Scudamore, the President of the Royal College of Veterinary Surgeons, the President of the British Veterinary Association, Ben Gill and Jim Walker, President of the Scottish NFU and *Channel 4 News*. It was headed 'The Slaughter of Healthy Animals'.

They wrote:

> We are veterinary surgeons who have been working for the past weeks to assist in the eradication of foot-and-mouth disease from our animals. It is vital to identify diseased animals at the earliest opportunity so that the spread of the infection to clean uninfected herds can be avoided. From time to time it is necessary to slaughter healthy flocks and herds, where they present a serious risk of becoming diseased.
>
> Unfortunately, the attempts to control this infection have been taken from the hands of the Chief Veterinary Officer and control now rests in your Office: we understand from the media that you are advised by the Government Chief Scientist and the Professor of Epidemiology from Imperial College, London University. It should be pointed out that neither of these men have any veterinary training. As a result we are now seeing a savage attack on what livestock remains in the north of

England and south-west Scotland. Animals are being slaughtered without rhyme or reason, often weeks after the supposed danger farm has been eliminated. This 'scorched earth' policy will undoubtedly result in the eradication of foot-and-mouth disease but it may be a pyrrhic victory.

Healthy animals will have to be slaughtered but the decision on their fate should be determined by veterinary surgeons, not politicians, soldiers or mathematicians. Our training was to save animal lives and to improve their health and welfare: veterinary surgeons have the unique privilege of being able to kill animals to prevent suffering, but this does not give us the right to indulge in mass destruction of healthy animals because of unproven computer predictions. If we are not to be used as professionals, we must question why we are employed at all.

What Blair's answer was is not known, but it was at that point that the policy was 'tweaked', and vaccination was again being mooted. As early as 30 March the government had applied for protective emergency vaccination for cattle in Devon and Cumbria and permission was granted, later extended to Cornwall, Somerset and Dorset, but the option was never taken up. A number of sources suggest that Blair was, in fact, in favour of selective vaccination in spite of the messages he was receiving from the NFU, but apparently, just before he called the election, he was presented with figures (massaged or not) and assured that the epidemic was now tailing off.

There is now a very substantial body of evidence to suggest that such a massive slaughter policy was wrong although it is still being implemented in and around the hot spots which are refusing to go away. The basic premise of the criticism is that if this really was the only route, the one policy that could possibly eradicate the disease without a colossal slaughter programme, a number of factors had to be in place. The first of these is absolute certainty as to the source of the first outbreak. The Waughs at Heddon-on-the-Wall are still the official prime suspects, but there is now a growing body of professional opinion to the effect that the Heddon outbreak was not the first, that the virus might well have been present in sheep in Dumfries and Galloway from early December, as instanced by clusters of abortions among sheep in the area. Such abortions could be brought about by other conditions but there are vets who claim that they had found old lesions in animals well before February.

This, coupled with suggestions that it might have also been present in Heddon in December and reports from timber merchants and farm suppliers that they had received requests from MAFF for information on timber and disinfectant availability in early February, is interesting, although it is claimed these enquiries were merely 'routine'. But if this really was the case, then it raises the possibility that there were suspicions that the foot-and-mouth virus was already in the country before the first case was confirmed.

Richard North, a public health specialist and epidemiologist also working on agriculture and food safety, has been a constant critic. In an interview with Jane Mayes in July he stated that even if the authorities truly believed the epidemic to have begun with the Waughs' pigs, then they should have shut down all the marts, the livestock markets, nationwide immediately, despite the logistical difficulties, and then slaughtered all animals in marts with traceable contacts.

MAFF's District Manager for Newcastle, Andrew Hayward, when tackled on this point and also asked why, after all exports had been stopped, and Longtown identified as a link to Heddon-on-the-Wall, they did not put a banning order on movements from all markets, said that it would have been impossible to do this as there are 'tens of thousands of animals in the marts at any one time'. Also, unofficial movements take place around them. There simply would not have been sufficient facilities for feeding the livestock while awaiting slaughter. In contrast, Peter Greenhill, chairman of Mitchell's Auction Mart, the oldest in the country, says that, though it would have been very difficult to keep the animals in the marts, it would not have been impossible.

The result of this failure, as the critics point out, is that from the very beginning the slaughter policy had a great big hole right in its centre –

what is the point of chasing after the virus with the slaughter policy, when the animals had already been let out?

It was pointed out to Mr Hayward that since one of the major purposes of MAFF was to deal with major notifiable diseases, it is therefore obvious that MAFF should have kept its contingency plans up to date and have understood the way animals movements function in this day and age (which Hayward claimed that they did), and recognised that a major point of distribution of the virus was bound to be the markets. Staff should then have been sent in immediately to slaughter all livestock on welfare grounds before the animals spread it throughout the country. Hayward agreed, but said they simply did not have the resources in place to do it.

Anthony Gibson likens what happened to a forest fire. 'What we were faced with from the start was a multi-focused outbreak. It had got behind our lines before we knew it. It was like one of those forest fires which spring up in 30 different places with people frantically running around with fire blankets trying to bank it down, only to discover it's broken out somewhere else.'

To return again to Richard North, he concludes that the way the outbreak was handled initially makes nonsense of the subsequent contiguous cull policy. It has no sense and cannot be justified if there were neither the resources, nor the intention, to catch it at the first point of identification for, by the time it was implemented, it was already far too late.

But in spite of this, the views of most critics have been summarily dismissed and, in one instance, greeted with childish contempt. Before he finally left Britain for a top post in Canada, Dr Paul Kitching of the Institute of Animal Health in Pirbright, and an acknowledged world expert on the virus, was 'consulted' by the government as to his views on the policy of mass slaughter. But it seems he did not come up with the correct answer. When he was critical of the scientific modelling which backed up the cull, his concerns were rejected out of hand. As he told *Channel 4 News* on 1 June:

> They considered me a bit of a Neanderthal for not
> appreciating models, and questioned whether I even went

on an aeroplane because 'someone as primitive' as myself wouldn't trust an aeroplane. I told them that there hadn't been the time and resources to collect even the basic epidemiological information on this current outbreak which would be required as an input into such a (computer) model. The result was that this could provide totally misleading information. Again it went back to the origins of the epidemic and at the very beginning the virus was in place, it wasn't spreading significantly, it wasn't going anywhere, and if the resources had been available at the time, all these (animals) could have been identified and slaughtered.

Another academic whose views were discounted is Professor Fred Brown, a former deputy director at Pirbright who now advises the United States government on foot-and-mouth. He declared that the government's reluctance to vaccinate in the early weeks of the outbreak had been based on 'old science', which failed to take into account the progress made in vaccine manufacture and testing procedures. The whole question of vaccination will be examined in the next chapter but it needs to be re-emphasised here that the main opposition to it remained the hierarchy of the NFU and its obsession with the 'disease-free status' of meat exports, coupled with a barrage of lies and misinformation from MAFF. In view of what has happened and the catastrophic effect on British farming, the opposition of Ben Gill looks positively surreal. Who, since BSE, is clamouring to buy our exported meat?

Professor Brown recognised that the government's policy was driven by fear of losing disease-free status, but that modern techniques made it possible to test whether or not meat had come from a vaccinated animal and that there was no health risk from eating vaccinated meat. He also claimed that as early as 9 March the US government had offered Britain special hand-held testing equipment which would have allowed on-farm testing for foot-and-mouth within two or three hours – thus quite possibly saving hundreds of thousands of animals from slaughter – but this had been rejected by MAFF. Officials had told him that they were 'too busy' to deal with his offer.

Support for his position on vaccination also came from Dutch scientist Simon Baretling, one of the architects of Europe's 'disease-free status' policy who openly admitted that the very policy he had helped to devise had now gone too far, and was preventing vaccination even in the face of unprecedented outbreaks of the disease. He also gave the grim warning that there was now a serious danger the virus would survive through the summer of 2001 and break out again in the autumn. Other foreign scientists described Britain's response to the epidemic as 'mediaeval'.

Worst of all is the very real possibility that even if the contiguous cull policy was justified, over two million animals at the very least have been slaughtered needlessly. There are now some signs of official back peddling, of former proponents of the policy hinting that they had not altogether believed it had been necessary, that there really was no alternative. The fiercest critics point out that, taken to its logical conclusion, the only way the policy could be proved to have completely succeeded, was when every sentient being other than man has been culled from the British Isles.

Dr Paul Kitching remains adamant: the mass contiguous cull policy defied all reason, was 'a total suspension of common sense'. Anthony Gibson goes along with that. If the epidemic really has been brought under control – and the word is 'if' – then it has succeeded only through a terrible bloodbath. We are presently being assured that the disease is 'tailing off', though few figures appear in the press. But as a North Country vet, David Black (who has been working for MAFF on the outbreak in Cumbria and the Borders) commented at a press conference on biosecurity given by Junior Minister Elliot Morley in Penrith on 7 July, the current 'five to six outbreaks a day is not a tail-off, it just looks like it in comparison with what went before'.

What is truly appalling is that there are grounds to believe that the wrong diagnosis was made on 25 per cent of farms. On 27 June, during a long interview with the author, Anthony Gibson said that 60 per cent of the outbreaks 'confirmed' in sheep, later came back negative. A substantial proportion of so-called infected farms had never ever had the disease.

His comments were to be substantiated on 9 July when DEFRA

released new figures. Mistakes in diagnosis, the Department admitted, had led to the unnecessary slaughter of well over 200,000 animals. The new figures prove that nearly 600 farms suffered a contiguous cull when tests later showed that foot-and-mouth had never been present on the original supposedly infected farm. Devon, host to 170,000-plus rotting dead animals, suffered most. 50,000 animals had been needlessly killed. A total of 122 farms in the county had lost all their livestock, some 9,551 head of cattle and 44,176 sheep, where foot-and-mouth cases had been suspected, but later disproved. One wonders if the animals of the Winslades and Willmetts were among them.

Junior Minister Elliot Morley immediately leapt in again to defend the increasingly discredited policy. 'A negative test,' he asserted, 'does not necessarily mean that the disease was not present.' Nationally there were 598 similar cases, involving the slaughter of 35,432 cattle and 173,130 sheep. Were Carolyn Hoffe's five pet sheep among them?

Jeremy Roe, a West Country barrister and businessman writing of the policy of contiguous cull puts it bluntly: 'Foot-and-mouth is a non-fatal animal disease. Foot-and-mouth is not an animal welfare issue. (You don't kill 3.5 million healthy animals for reasons of welfare.) Foot-and-mouth is an economic problem for the farming industry. It's a problem but only an economic problem for one industry within the rural economy. Foot-and-mouth didn't kill four million of our farm animals. It didn't leave thousands of carcasses lying in farmyards and fields. It didn't bulldoze thousands of carcasses into vast open pits in our countryside. It didn't burn thousands of carcasses in open pyres near our towns and villages. It was the government's response to foot-and-mouth that did all those things. Foot-and-mouth did none of them.'

The reason for the government's actions was simple. It was done to 'protect the farming industry against losses which were only a fraction of those suffered by every other business in the rural economy. To protect the farming industry against the economic consequences of a non-fatal disease, billions of pounds of damage was inflicted to the rest of the economy. Of course the outbreak had to be dealt with but there were other remedies available, remedies that have been used by other governments. And given what has happened, who believes that the farmers have been helped by government's actions?'

Commenting on the fact that so many farms were found never even to have had the disease, Anthony Gibson has said that MAFF had simply decided it couldn't afford to take a risk. 'If there was the slightest doubt, then kill the animals, kill the neighbouring animals, kill everything in sight. The contiguous cull is one of the most bloody, tragic, and disgraceful misjudgements ever committed.'

# Chapter 10

# Vaccination – Lies, Damned Lies and Statistics

The background to it is the strong prejudice in the veterinary establishment against vaccination. Not just government vets, but academic and private practice vets as well, based on historical factors. We've always looked down on those countries forced to vaccinate. Looked askance at those countries on the continent as 'useless French and Dutch that can't cope with a foot-and-mouth epidemic without resorting to vaccination'. So people didn't have open minds when it came to considering it as an option. Most people had already made them up, based on experience which was 35 years out of date.

This included the government's own veterinary advisers who told us all sorts of lies about what vaccination involved. I promulgated all this myself too, as I'd been told it by my Head Office and by the Ministry of Agriculture as well: statements such as all vaccinated animals still having to be killed. That was the policy, so I duly wrote it in our bulletin and in my column in the

*Western Morning News.* I then managed to get hold of the Standing Veterinary Committee only to find they'd said nothing of the sort. So I had to apologise to my readers.

We were then told that, yes, vaccinated animals could live but they'd have to be kept under cover for 12 months. I was told that by my own Director. Told it by the Chief Vet. When you're told that kind of thing from those people, then you tend to believe it. So we put that in our bulletin and all the farmers immediately threw up their hands and said 'God, that means a twelve-month winter, we can't cope with that. W haven't got the facilities.' So farming opinion, which had always been suspicious of vaccination because it was 'what third world countries did', swung heavily against it; the deception encouraged by the establishment to believe that if we vaccinated our livestock would be blighted forever.

Nor did the deception stop there. We were told there were no tests that distinguished between antibodies caused by vaccination and those caused as a result of the disease. That what there was had not been tested properly. Therefore, by the time they actually started being honest, about early April, it was too late. The damage was done. Farmers' opinion, because their questions had not been answered, began to turn against it. At the outset those in favour and those against stood at about 50/50 – pedigree people for it, commercial chaps against. For commercial farmers the compensation rates are good, so there's a vested interest in slaughter. But later opinion swung 70/30 against. Anyway, by that time it was operationally too late.

A bitter Anthony Gibson, South-west Regional Director of the NFU.

The story of the opposition to vaccination is one of ignorance, mindset, vested interests, fear of losing face, chicanery, deception and downright lies. It is not a pretty one. Nor is it in any way irrelevant since, at the time of writing, new and unexpected outbreaks in

Yorkshire are prompting fears of a spread down to the vast pig farms of East Anglia, rightly described as a nightmare.

As has been pointed out previously, pressure for the use of vaccination began very early on in the epidemic but from the first it has come up against a wall of prejudice. During that crucial period in early April, when so much might have been achieved, several high-powered commentators were forcibly putting the case forward for vaccination in a variety of media. One of these was Jonathan Dimbleby, President of the Soil Association and an organic farmer as well as a journalist. In a feature in the *Daily Mail* on 6 April, later circulated by the Soil Association, he pointed out that the organisation had qualms about the slaughter policy from the start and so had gone public with a strategy which offered the nation 'the only available prospect of an early reprieve from the mass carnage' and had continued urging it for the last three weeks, even if it were only used as a 'firebreak' in the hot spots of Cumbria and Devon.

The reluctant response from government was that it might possibly be used as a 'last resort', but Dimbleby points out that the 'first resort', even then, had cost the deaths of over a million animals. 'The ballooning cost of compensation to farmers alone is expected to drain the Treasury of a billion pounds.' (This has turned out to be a major underestimate.) 'While most of the countryside remains firmly "closed for business", the financial loss to allied enterprises in the rural economy and to the tourist trade is already a calamity. No one knows how long it will be before the disease is eliminated, or what the eventual cost will be. The Treasury talks about £3 billion – equivalent to a penny on income tax. Others foresee losses to the economy of up to £9 billion.' (Recent analysts have even suggested double that figure.)

He goes on to say that this is the price of putting in place a scorched earth policy to re-establish an export industry in live animals that is itself built around a crazy system of Common Agricultural Policy, subsidies that indirectly sustain agribusiness attitudes, which turn sheep into commodities and dealers into brokers 'shovelling the wretched creatures around the country from one market to another before shipping them across the channel to be slaughtered in Normandy and sold in Paris supermarkets as French lamb'.

He then goes for the jugular. 'But vaccination is not only a "last resort": it is vigorously opposed. The Ministry of Agriculture and the National Farmers' Union have formed an unholy alliance which appears to have secured an arm lock on government policy. Yet they have only achieved a pyrrhic victory by arming themselves with a quiver-full of myths in which they may sincerely believe but which fly in the face of scientific advice.'

The most basic of these myths are those also peddled to Anthony Gibson: that vaccination is ineffective; that vaccinated animals could not be sold for human consumption; that milk and meat from vaccinated animals could not be sold for human consumption; and that vaccination would unnecessarily postpone the date at which Britain would regain its 'disease free status', without which no animals or meat exports could leave these shores. Even a cursory reading of the available evidence from Europe and the US Department of Agriculture would show quite clearly that all these claims were either false or thoroughly misleading.

'In my part of the world,' Dimbleby concludes, 'a lot of farmers are also infuriated by the metropolitan assumption that the NFU always speaks on their behalf – even when, as in this case, its principal motive appears to be to ensure that the government showers its stricken members with financial compensation on an unprecedented scale and never turns off the tap.'

Dimbleby is one of many critics suggesting that the hierarchy of the NFU has had its own reasons for holding out against vaccination. When it looked as if it might happen, Ben Gill produced his 'list of questions' – some 51 in all – all of which had to be answered before it could even be considered, a tactic which is reminiscent of the endless impasse in Northern Ireland and one which must have delighted the officials at MAFF.

Magnus Linklater in *The Times* took a similar critical line at a later date, pointing out how the election had knocked the true state of the epidemic out of the news; that no one was now telling us how many animals were being killed and 'worse, government policy is based on bad science'. Those seeking to challenge government data have been 'routinely sidelined', co-operation refused to those who have offered to help with testing and independent expertise rejected. He points out that we will probably never know how many animals really did have the disease since most of those killed 'on suspicion' or as 'dangerous contacts' are no longer tested, 'which is not only sloppy procedure but rank bad science'. The picture that emerges is 'of a government machine driven by ideology rather than expertise. Determined to defeat the disease by the sheer weight of the slaughter policy it has imposed, it seems incapable of absorbing up-to-date research from anyone outside the magic circle that might challenge its credentials. Since it has the farming union on its side, it can ignore well-founded arguments for vaccination that would have saved the country billions of pounds, rescued its tourist industry and prevented the needless slaughter of healthy animals.'

This kind of opposition to government policy leads us, once again, to the fact that instead of taking advice from our own world authority on the subject, Dr Paul Kitching, before he left for Canada in despair, or a previous member of the Pirbright team, now adviser to the USA on the subject, both international experts were contemptuously dismissed in favour of a scientist whose expertise lies in metals, and a team of epidemiologists and computer modellers without any veterinary expertise at all. There is no shortage of academic and practical research material available, all pointing in the same direction.

First, therefore, to examine the purely economic case for vaccination as opposed to slaughter, there is an interesting paper, 'The 2001 FMD outbreak: Economic arguments against the cull' by Peter Midmore, a professor at the Institute of Rural Studies, the University of Wales, Aberystwyth. 'It is the intention of this briefing paper,' he writes, 'to demonstrate that, whether or not these measures (i.e. the contiguous cull) bring the outbreak under control, their cost would exceed the main benefit secured from not adopting the alternative strategy of vaccination. The argument, by the government and the main farming unions, has been that because vaccination makes it impossible to distinguish between healthy and infected animals the UK would lose disease-free status, and exports of livestock products to third countries could legitimately be banned.'

On the one hand, he posits, is a loss of export markets, which implies a loss of revenue and the cost of prophylactic vaccination. On the other hand, the extended cull will require compensation for all livestock destroyed; the costs of disposal; consequential losses for slaughter farms due to loss of sales and the quarantine period; consequential losses for hauliers, processors, other multiple impacts and the effects on tourism.

He then goes into a series of arguments, with tables based on MAFF's own figures, showing that the major farming exports are 'sheep meat and dairy products', followed by beef, pork, and poultry. While the overall meat and dairy export market in 1999 was £705 million, the only exports which would now be affected would be lamb, beef and pork, valued in 2000 at some £310 million. He then goes through a detailed and lengthy process of analysing the costs of both policies, the number of animals then involved in the extended cull, the consequential loss of livestock and business to farmers, the rest of the rural economy and to tourism, set against the cost of vaccination (which he puts at approximately £15 million) and the possible loss of export markets, and concludes that the economic arguments are against the policy of extended cull in almost every aspect.

Dr Ruth Watkins' pro-vaccination paper dismisses at the outset the argument against the extended cull as a means of control.

> A proper and comprehensive vaccination programme raises protective antibodies if it is applied to a whole herd, which not only attains control but would also eliminate the virus from the country with only minimal slaughter. The train of infection from one acutely infected animal to another susceptible animal can continue indefinitely where there are large populations. New-borns add continuously to the susceptible population each year. Vaccination of the whole herd effectively gives the virus nowhere to go so it dies out.
>
> The virus is hardy and susceptible to spread in many different ways and survives for a long time in the environment. The present PanAsian strain, the O-serotype, has been known for ten years and I have every

reason to believe that as with other members of the picornaviridae, a vaccine similar to the infecting serotype would be effective, as in polio virus infection in humans. This would appear to be the case as large stocks have been prepared of suitable potent vaccines and there is experimental evidence and experience of their use and proved efficacy . . . therefore the vaccines ought to be effective and government vets would know this.

The vaccine, she continues, is in the form of a killed virus, 'in effect a shot of protein'. It can be given to any animal, pregnant or not. Like human infants, young animals too could be vaccinated. This would be important to attain a herd immunity of at least 80 per cent. This is the level required in foot-and-mouth disease to control and eliminate infection. The immune response of the potent vaccine prepared is significant after even one dose of vaccine. A low titre of antibody would be fully protective and the protective level of antibody may be reached in about 4–14 days after the first dose of vaccine. 'New infections should cease three weeks after the population is vaccinated. Thus vaccination can be used to damp down infection in already infected flocks by preventing later cycles of infection.' It could also, she points out, be used to vaccinate all stock within a 50-mile radius of a source of infection, then rolled out to include the whole country.

The essential point, she emphasises, is that vaccination should be done as swiftly and completely as possible. 'It would be much quicker and easier than the current policy of diagnosis, slaughter, disposal and decontamination. One could stop this and put all resources into a vaccination programme. I would vaccinate all animals regardless of status, infected or not infected, incubating disease or recovered, beginning with the infection regions or counties.'

She points out there are examples of human diseases where vaccination has been essential in their control, for instance measles spread by the aerosol route and polio virus spread by respiratory secretions and in faeces. In the case of the polio virus, many infected persons shed the virus in faeces for up to six months but this has not prevented the success of vaccination. The polio virus is now eliminated from both North and South America.

She concludes by saying that the foot-and-mouth disease vaccine has been used extensively, preventing epidemics recurring in Europe; in Uruguay to rid the country of endemic foot-and-mouth disease; and as a control of outbreaks in hot dry countries where the domestic animal population may be sparser than our own. Logistically it would be quite possible to vaccinate the animal population, especially if the army were called in to help with the distribution. 'We are in the 21st century. MAFF seems caught in a time warp of 40 years ago.'

Further back-up comes from Abigail Woods of the Wellcome Unit, Manchester University, in her paper 'Evaluating the Current Control Policy from a Historical Perspective', referred to earlier in the book, also 'Foot-and-Mouth Disease in the UK: Problems with Current Policy and Feasibility of Alternatives' by Dr Keith Sumption, a veterinarian with a PhD in molecular virology, from the Centre for Tropical Veterinary Medicine at Edinburgh University, and in a series of counters to statements made by the Chief Vet and Ben Gill.

Abigail Woods points out that when large-scale slaughter has been used and yet the disease continues to spread, 'opposition has frequently been directed to the sheer scale of the destruction. The Ministry tends to counteract this by stating that the percentage of livestock killed in national terms is extremely low. This is an attempt to disguise the fact that in certain regions, percentages are huge – 33 per cent of Cheshire cattle in 1923–24 and 1967–68. In these cases, farmers argued, that disease controls had failed and that elimination only occurred because there were no longer any livestock left to infect.' She too agrees that historically the NFU executive has always supported MAFF in the decision to slaughter. 'However, at grass-roots levels, there has been considerable dissent, but regional opinions are often discarded by HQ.' .She supports this assertion by pointing to NFU minutes from the meetings of the Meat and Livestock Committee held at Reading Rural History Unit. She notes that members of the medical profession have historically criticised the slaughter policy. 'For obvious reasons, doctors tend to rely on therapy and vaccination for disease control and this reliance on scientific, laboratory-formulated measures, has shaped criticisms of a supposedly backward and barbaric slaughter policy. However, medical criticisms have been repeatedly rejected by farmers and vets on the basis that doctors are only experts in the field of human disease and have no role to play in the management of livestock problems. It is important not to overlook the fact that certainly prior to the Second World War, medics and vets were competing for "territory" in terms of which profession should be responsible for milk/meat inspection, and for research into animal diseases.' Opposition to slaughter therefore tends to be written out of history, 'precisely because the individuals concerned are not always the most prominent or influential'.

So much has changed since the 1967–68 outbreak, not least the size of farms, the smaller number of people involved in farming and the sheer numbers of livestock involved. Not only has the whole context for foot-and-mouth control changed, but also the disease itself must be reinterpreted. The belief that Britain can eliminate the virus by slaughter has meant that while publicly MAFF expressed hopes that a vaccine would emerge from Pirbright, in private it was clear that vaccines would not find any application in Britain.

'The fact that FMD is such a contagious virus justified the restriction of research, at least on large animals, to workers employed at Pirbright under the FMD Research Committee, over which MAFF had a huge degree of influence. This made it impossible for researchers to investigate the disease and formulate alternative methods for its control.' However, while many advances have been made in vaccination, it is clear 'that to MAFF these advantages will never be sufficient. The Ministry keeps moving the goal-posts, such that nothing short of no-risk, 100 per cent protection will be regarded as sufficient. This could hardly be claimed of any vaccine in existence. While good progress has been made in tests to differentiate infected and vaccinated animals – tests that have important implications for the export trade and considerably strengthen the case for vaccination – MAFF rejects these as insufficiently advanced for field application.' This again, Woods points out, is a long-standing argument but no aspect of vaccine technology has ever been sufficiently advanced for field application, according to MAFF.

Other nations, currently disease-free, are far more open to vaccination and not all European nations were happy at the decision to stop vaccinating against foot-and-mouth disease within the European Union in 1991, a decision heavily influenced by Britain 'in order to streamline disease control policies and lift trade barriers, as recent comments in the press suggest'.

She concludes that MAFF has set completely unobtainable scientific criteria that supposedly justify its rejection of vaccination but actually just support its pre-existing decision not to vaccinate; that the true reasons for not vaccinating are grounded in misplaced confidence; that because slaughter has always worked in Britain, it will again; and then there is the matter of national pride. MAFF feels that Britain is superior to vaccination. Finally, if MAFF did turn to vaccination now, when a decision could so easily have been made earlier without the slaughter, it would seriously undermine its legitimacy. 'Also, in a sense, this would betray all those past CVOs who put their careers on the line to withstand farmers' complaints and assert that slaughter was the best and only way to control FMD. The fact that the slaughter of up to a million [now over three million] animals is supposedly justified to save a single government department's credibility can surely not be tolerated.'

Dr Sumption's well-argued paper covers much of the same ground as those previously cited, but goes into great detail and emphasises that this particular epidemic is outside our experience. 'Therefore we are in uncharted territory in the use of control by slaughter for national or international foot-and-mouth disease control, without precedent or example to guide. In contrast a combined stamping-out and emergency vaccination policy has often been used on a large scale, most recently in Algeria in 1999, where 165 widely dispersed outbreaks with a type O virus occurred, but were controlled within six weeks of the start, with emergency ring vaccination commencing in the first week, with over two million vaccinations and the greatest reduction within a month of the first case.'

He becomes even more forthright when answering some of the statements made by Chief Vet Jim Scudamore and Ben Gill, on the Sheepdrove website of which a few examples will suffice.

Statement: First of all you need multiple injections, a primary injection, a booster after one month and then every six months.

Response: A travesty: emergency high potency vaccine can be used alongside slaughter. In Algeria in 1999, the outbreak was brought under control before the need came for the booster and the vast majority of cattle only needed the one injection.

Statement: It's not effective for at least 15 days.

Response: Not true. High potency vaccine offers substantial protection after 2–4 days. Assume 7–14 days for full immunity but in both pigs and cattle the vaccine reduces the 'titre' and duration of virus excretion.

Statement: Animals still have to be slaughtered.

Response: Scientifically validated tests exist in Europe, South America and the USA, which can distinguish between antibodies acquired from vaccination and those resulting from infection. Intervet announced commercial tests on the day Ben Gill said this.

Statement: Vaccine requires repeat injections.

Reponse: This is not correct. High payload emergency vaccines held at Pirbright need only one injection to confer protection from disease and virus shedding. 500,000 doses of suitable vaccine are available.

Statement: It's not 100 per cent effective.

Response: There has never been a 100 per cent success rate in a vaccine for anything, yet their role in assisting disease control is universally accepted.

Magnus Linklater, writing in *The Times* on 24 May, cited another area of specialisation to which MAFF's approach seems cavalier – 'understanding basic virology'. 'Recently,' he said, 'one UK virologist used by the NHS and the Public Health Laboratory to test blood for HIV . . . contacted MAFF. He wanted to obtain some non-infectious treated virus or extracted nucleic acid to create a sensitive diagnostic "assay" – a means of detecting infection in the blood and thus making an accurate early diagnosis of foot-and-mouth disease. It is a rapid system which MAFF does want not to use, preferring to rely on old-fashioned labour-intensive tissue culture methods taking several days. Again the approach was turned down. The virologist was told that if he attempted to obtain potentially infected specimens independently, he would commit a criminal offence. What shocked him, however, was an apparently casual remark by a senior MAFF veterinarian to whom he spoke, which suggested that the MAFF man did not know the difference between a protein and nucleic acid – a fundamental misunderstanding which raises serious questions about MAFF's expertise in this vital area. None of those in senior positions advising the government have any known expertise in virology.'

Vaccination may not be the 100 per cent perfect solution but what are we now to make of the policy of contiguous cull? There is now sufficient hard evidence to lead to properly informed debate but there are alarming signs that hard evidence backed up by research will still not be sufficient. Ben Gill's mindset remains total. After hearing the arguments in favour of vaccination rehearsed yet again, his response, quote in the *Independent on Sunday* was, 'Bilge!'

Since these papers were published in the spring of 2001, MAFF has become DEFRA. Even if the government had not thought that an umbrella department covering the environment, food and agriculture would be a good idea, along with the departure of the luckless Nick Brown, the name MAFF had become too much of an embarrassment. So it was with a certain amount of fanfare that MAFF became the Department of the Environment, Food and Rural Affairs with its brand new Minister, Margaret Beckett. Elliott Morley drew the short straw which put him in charge of livestock in general and the foot-and-mouth epidemic in particular.

But simply changing the name in an attempt to rehabilitate an organisation is not enough, as instanced by Windscale, British Nuclear Fuels notorious reprocessing plant on the Cumbrian coast, becoming Sellafield. 'If it looks like a MAFF, smells like a MAFF and sounds like a MAFF, it's a MAFF,' wrote one correspondent to the *Western Morning News*.

By the middle of July there was a new, and totally unexpected, outbreak of the disease centred on Thirsk in Yorkshire. On 11 July, senior vet Fred Landeg announced that the government had issued a red alert over the threat of a North Yorkshire outbreak spreading to the region's intensive pig farms (only ten miles away) and beyond into Lincolnshire. Mr Landeg, described as 'the head of the veterinary exotic disease team at DEFRA', urged farmers in the Thirsk area to take all possible biosecurity precautions, warning that pig farms are 'virus factories', because they 'breathe out vast quantities' of the foot-and-mouth virus. The Thirsk outbreak could, therefore, be very serious indeed.

On 12 July the airwaves were filled with infuriated farmers. DEFRA vets, mainly foreign and from the continent, were wandering around the area totally lost, blundering into farms to find out if that was where they were supposed to be culling livestock. Officials in four-by-fours were racing round the lanes, reversing into farm gates before going on somewhere else. On one occasion a convoy of officials had gone down a gated road and driven through a flock of sheep. What was needed, pleaded the interviewees, was some efficient, local, overall control based in one place.

A senior DEFRA vet, Fred Landeg, was hauled on to the *World at One* that day to give his views on these worrying incidents. Presumably this was the very same Mr Landeg who had issued a red alert around Thirsk and urged farmers to take all possible biosecurity precautions. He began by stating that he did not think the incidents he had just heard described had actually happened.

'Are you suggesting these guys are making it up?' asked the interviewer. There was a pause. No, replied Mr Landeg, cautiously, but he had not heard of these problems although he would look into them if he was officially informed. Then he came up with the message now constantly being peddled by DEFRA when asked about the progress of the epidemic and fears of further spread. Stopping the disease is now down to the farmers. Not government, not DEFRA, not vaccination, but the farmers and their biosecurity. Had farmers properly shut their gates, he pointed out, vehicles could not have reversed into them so there would have been no risk. Pressed, however, on the incident when a MAFF team went down a gated road, opening gates as they went as well as driving through a flock of sheep, Landeg went into reverse. If the matter was sufficiently urgent then obviously officials might have to open the gates on gated roads. As for people asking for local control, there were people there seeing to things even if they did not have a specific office. As one farmer's wife commented, when asked if she had noticed a difference since the setting up of the new department, she replied there had been a new nameplate on the door and new headed notepaper but that, so far as she could see, was about it.

At the time of writing there are very real fears that the Thirsk outbreak could indeed spread to the massive pig units further south. But from enquiries made by the author, although there have been rumours of a change of attitude towards vaccination, there is no official sign of it. Once again it is off the agenda.

The single most shocking comment was made by Anthony Gibson. From first to last we have been told that it would take 12 months after vaccination before we could resume meat exports, the main plank of NFU antagonism. Keith Sumption suggests that 12 months is actually the worst-case scenario. But it seems that, if the main reason for the authorities' attitude towards vaccination (apart from national pride at 'disease-free status'), is the effect it would have on our meat exports, then opposition to it is irrelevant. 'Because,' says Gibson, 'the law regarding the resumption of exports is *exactly the same* [my italics] as that for slaughter. The prohibition on both is six months. Six months after the last outbreak or six months after the last animals have been vaccinated, you can resume exports.'

# Chapter 11

# Civil Liberties and Gun Law

*'Necessity is the plea for every infringement of human freedom. It is the argument of tyrants; it is the creed of slaves.'*

WILLIAM PITT THE YOUNGER, 1783

Among the many issues that will need to be looked into and discussed when the epidemic is finally at an end, is the question of civil liberties and what happened to them during the course of the outbreak. MAFF, as it then was, had almost unprecedented powers to go on to land. Rules, both European and national, have been flouted on a range of issues from pollution to slaughter. Then there has been the use of the Official Secrets Act.

Even to those used to the Act being dragged in to cover government embarrassment, it came as something of a surprise to learn that it was being used by MAFF/DEFRA to silence farmers who might want to talk openly about what had been going on. This interesting fact was brought to public notice when it was reported that the Waugh brothers were going to be taken to court for the state of their pig unit and also for failing to report the presence of disease. Bobby Waugh, as well as denying that his unit was the source of the entire outbreak, told reporters he had refused to be bound by the Act as he would otherwise be unable to mount a proper defence.

THE YEAR OF THE PYRES

From the beginning he has accused the government of making him a scapegoat and revealed that for 100 days after the disease was traced to his farm, MAFF refused to allow him to clean up his premises, although usually such cleansing was carried out within days of animals having been culled. He claimed that the refusal had cost him £24,000 in wages, rent and lost income. MAFF, then DEFRA, had told him he could only clean up his farm if he signed the Official Secrets Act. A leaked 12-page MAFF contract, dated 4 June (three days before the election), details the agreement which would enable Mr Waugh to claim £10,000 in compensation for the clean-up. But before he could even begin the necessary procedures, he had to agree to abide by Clause 17 of the contract.

This states that: 'The supplier undertakes to abide and procure that his employees abide by the provision of the Official Secrets Act 1911 to 1989. The supplier shall keep secret and not disclose any information of a confidential nature obtained by him by reason of the contract, except information which is in the public domain otherwise than by breach of this provision.' The contract would have been a lifetime gag on Mr Waugh. 'The provision of this Condition 17 shall apply during the continuance of this Contract and after its termination, howsoever arising.'

Why, it was asked, was it necessary to shroud what happened on farms in the mantle of state secrecy? How could anything whatsoever connected with the epidemic be considered so sensitive it might harm the security of the state? In opposition the Labour Party was all for a complete rewrite of the old and much discredited 1911 Official Secrets Act with its catch-all Section Two. It was considered a joke that Britain was so in love with the concept of secrecy that just about anything from the number of NHS beds recently taken out of use, through the safety (or not) of food colourings, to whether or not your local railway line was safe from closure, was Officially Secret.

It was generally agreed that secrecy had become a habit, a way of thinking, an attitude of the official mind noticeably prevalent in the British civil service. The Official Secrets Act was a very convenient tool of government for it could cover up stupidity, ineptitude, inefficiency, bad decision-making, bureaucratic bungling, dishonesty and outright corruption. It provided governments, civil servants in ministerial departments, office holders, businesses and any other organisation or bod, with a comfortable defence against having to be accountable to the mass of ordinary men and women. It seemed obvious that it had to be repealed and any subsequent legislation brought up to date.

Nor was that all. Alongside a new Official Secrets Act must come its twin – a Freedom of Information Act. But behold! Once New Labour achieved office, enthusiasm for change rapidly waned. The most obvious sign of this was when David Clarke, the Minister put in charge of drafting the new Freedom of Information Act and who was, it seems, making an excellent job of it, suddenly discovered that he was being viciously briefed against by shadowy figures from within the party, some at a high level in the Home Office, others not unadjacent to 10 Downing Street. The result was that he lost his job in the next re-shuffle and a watered-down version of a Freedom of Information Act finally crawled through parliament, after a great deal of opposition.

So, once again, official secrecy has been trundled out to save embarrassment and prevent people from learning what is going on. It might seem to everybody outside government circles and MAFF/DEFRA to be surreal, but it would seem that if farmers wanted their farms properly cleansed, disinfected, and finally cleared for the purchase of new livestock, then sign it they must. Asked why he thought this was happening, Anthony Gibson admitted that he had no idea. There are still a lot of things in MAFF/DEFRA's domain which are officially secret, he told the author, even though everybody knows about them. He instanced the boundary between favoured and less-favoured areas for government intervention. These are Officially Secret. 'You were allowed to look at a map showing them, but not take copies of it or talk about it!' He considers its present use to be a hangover from the past, a legislative anachronism.

He may be right but it has obviously proved a useful cover-all in the present circumstances. Will it be used if the government are finally forced into a public inquiry on the 2001 foot-and-mouth outbreak, one wonders? Yet there has been no protest against its use. Certainly nothing from parliament, but then who looks to the British Parliament to safeguard the rights of the individual today?

Waugh was warned that if he signed the Act it could prevent him publicly defending the charges brought against him by Northumberland Trading Standards Department. On 18 July Bobby Waugh was brought to court to face 16 charges, including five counts of failing to notify the authorities of foot-and-mouth disease in his animals, four of causing unnecessary suffering and one of bringing unprocessed catering waste on to his premises and feeding it to his pigs. Tynedale magistrates adjourned the case for a pre-trial review on 16 August.

The epidemic has also given the police, supporting MAFF/DEFRA officials, unprecedented rights to enter private property. When MAFF decided the Winslades' animals would be culled, police provided the MAFF vet who came to 'check on the animals', with an escort, came back again at 5 a.m. to see what was going on, and provided yet more back-up when the slaughter team eventually went in to cull the livestock. This is because, unless the circumstances are exceptional, the current foot-and-mouth regulations gave MAFF, and give DEFRA, almost unlimited powers.

Police have played a prominent part in forcing through culls where there have been protests, not least, as we have seen, in the case of Carolyn Hoffe's five sheep and the attempt to force a way into the Mossburn Animal Centre. But they have also appeared in force when there have been protests over pyres or pits on the public highway in Wales, notably in Wales where, as reported in Trecastle, one witness described police reaction as overkill, not least when it was discovered police had been drafted in from the Midlands as well as from other forces in Wales.

Yet this is the body that everlastingly complains that it has insufficient manpower and resources to be able to do its real job properly, i.e. catching criminals; and declares that it cannot stop football hooligans and other troublemakers roving the country. Though, again, they managed well enough during the miners' strike when it was almost impossible to drive through public roads in north Warwickshire (and no doubt elsewhere) without coming across roadblocks or police cordons on the lookout for flying pickets. But there seems to have been no problem pulling in hundreds of extra police to deal with foot-and-mouth protests. One can only wonder how big the bills will be when they are finally presented to the local authorities concerned.

As a correspondent to the FarmTalking website put it: 'We are witnessing the State coming on to people's property and saying in effect:"I am here to take your property, and after I've taken it I'm going to kill it. And if you try to stop me then I'm going to push you out of the way. And if you get angry I will arrest you." Indeed, if the State can march on to our property and destroy our life's work in this way, then we need to ask:"Do we really own anything?"'

There is no doubt that the police found themselves in a very difficult position, given the sweeping powers awarded to MAFF to carry out the contiguous cull and there are a number of instances of individual policemen getting upset at incidents they had to witness but were legally prevented from intervening. They were also having to deal with people in a highly emotional state, farmers and their families who were watching an entire life's work destroyed in front of their eyes, where many of the animals involved had been raised and cherished for years. Not to mention those who, in the lottery of MAFF's slaughter policy, found themselves on the losing side when their neighbour down the road was one of the winners. At the time of writing there are a number of threats of legal action over allegedly illegal entry and illegal culls. It is easy for feelings and tempers to run high on both sides.

But whatever else, the foot-and-mouth epidemic has done little for public–police relations in rural areas where, as a rule, the very sight of a policeman is a rarity. Whether or not the use of the police in such a manner will feature in any future inquiry, if there ever is one, is by no means certain.

Then there are the gunmen. Most, we know, were MAFF vets using bolt guns. The provenance of others, the gunmen taking pot-shots in Devon, Wales and Yorkshire, is more doubtful. This prompted a lengthy article on the Sheepdrove website written by Colonel Barry S. Turner (retired), under the title 'Marksmen and Gun Law'. In it he voiced the concerns of many.

He had been struck by the casual references to 'marksmen', apparently accepted without comment or reassurances, as he had hitherto been under the impression that there had been severe restrictions in place as to the ownership and handling of guns since the tragedy at Dunblane. 'Yet no one, least of all the NFU as the body that claims a national public profile, has uttered a squeak. Who are these marksmen who have materialised out of the woodwork so opportunely? Where do they come from to be allowed or "licensed" to run around like something out of a wild west show?'

The questions he goes on to raise require answers. As well as wanting to know who they are and their place of origin, what he asks are their qualifications? Are they regular members of gun clubs, policemen, servicemen, contract slaughtermen handed weapons temporarily? How is it possible to locate, organise and recruit marksmen in the numbers required? Exactly where are these weapons kept and what type of guns are they? Are they really suitable for the task? Where have they come from and in what quantities are they held for legitimate use and, if so, by whom? Are they on police registers? If so, under what justification? Are they licensed to individuals, gun clubs, competition sportsmen? Or are they all police owned?

And what about security? How are they secured during the extermination programmes? In locked cabinets? In a hotel or bed-and-breakfast accommodation being used by the marksmen and MAFF officials? In police custody between shootings? How are they transported to slaughter sites? Moved in public places by cars, vans, coaches? Are they issued by the police? If so, where? On site?

Colonel Turner is also interested in the ammunition and how it is stored and accounted for. That 2,000 rounds may be issued to one person at one time seems incredible. 'Can we assume, therefore, that ALL checks on marksmen are suitably qualified and accredited to hold arms in any quantity? What checks have been carried out to guarantee this? What action is taken to check unspent ammunition after shooting, check and clear weapons and check the area for lost ammunition?'

Lastly, there is the question of public safety and the precautions (if any) taken to protect them? Are warning signs put up or flags displayed? Is land cordoned off and the cordon enforced? What precautions exist against undisciplined behaviour by marksmen or to prevent injuries to farmers, members of the public or even other marksmen? Is this supervised by qualified policemen, army weapons instructors or who?

He concludes on a warning note. 'Regulations in armed forces both on operations and training are unequivocal and very specific. If this situation occurred in Northern Ireland, media and politicians would go berserk. Heads of senior officers and policemen would roll. But I have seen no public information, assurances or challenges from individuals or reputable bodies. It seems to imply the law has been thrown to the winds.'

# Chapter 12

# Pollution

It is not only questions about guns and public safety that remain unanswered. There are whole areas where answers to important queries are at best vague and at worst non-existent, not least those concerning what impact the epidemic and the mass cull might have on the environment. DEFRA refers the enquirer to its website and the section on 'health and safety', where the only items to be found are to do with biosecurity.

Also, somewhat surprisingly, the possible long-term environmental hazards of the effects of foot-and-mouth disease do not appear to have seriously interested the major environmental bodies such as Greenpeace and Friends of the Earth. Neither could answer questions as to the possibility of water courses becoming polluted, nor had they undertaken any research on the smoke from pyres or the constituents of the ash. For whatever reason they simply did not see this as coming within their remit. The Soil Association has so far undertaken no research, preferring to concentrate its fire on the battle for vaccination and the Environment Agency (EA), while sympathetic, has little or nothing to add to what has appeared in the press or in its own press releases.

The EA is a 'non-departmental body', that means it is a privatised part of government, which was 'sponsored' by the Department of the

Environment, Transport and the Regions until the general election, and now works in association with DEFRA. A certain proportion of its money comes from government, the rest from charging polluters for licences to pollute.

The one document published by the EA is on the subject of sheep dipping and the measures that need to be taken to avoid the virus being spread during, or after, dipping takes place. It is in the form of a guidance note on how to proceed during the current epidemic. Many farmers will be dipping or showering sheep in the coming months and 'it is essential they take the utmost care'. All sheep dip must now be treated to kill the foot-and-mouth virus and be disposed of carefully to avoid pollution. Unless the used dip is taken to a licensed landfill or incineration site, disposal should be undertaken by spreading it thinly on to suitable land. Preferably this should be at the farm where sheep are dipped. Concern is expressed that some farmers rely on mobile contractors to take used sheep dip away for disposal but DEFRA advises that contractors should not take used dip from inside an infected area and dispose of it outside that area unless it is going to a licensed landfill site or for incineration.

The Agency then gives the advice: 'If the sheep dip bath has a drain hole then it must be sealed; the bath used requires testing for leakage and should be filled with water and left overnight to check there is none; all spillages must be contained and it must be ensured that all drainage runs back into the dip bath. Sheep must be kept in drain pens until they have stopped dripping, and out of streams until their fleece is dry. Even after the dip is "treated" to kill the virus (with what is not specified), it is still "very polluting". If spread on land it must be diluted and it must not be spread on land which is steeply sloping, poorly drained, waterlogged, has cracked soil or which is important to wildlife. Nor must it be spread within ten metres of any watercourse or fifty metres of any well or spring.'

This all sounds fine, except that the question is raised as to how all this is to be monitored? Who will check that all these precautions are scrupulously being carried out? Anyone involved in past campaigns against overuse of pesticides will know of the countless examples of far larger amounts being used than recommended, and of cocktails of various chemicals being mixed together in the hope of obtaining a better result. This author, researching for a Channel 4 television documentary back in the 1980s, discovered that forestry workers who had been using 245T, one of the herbicides which went to make up the notorious Agent Orange used by the Americans in Vietnam, had regularly been washing out their equipment in a watercourse which fed directly into a reservoir, which was itself a main domestic water supply. What reason have we to suppose that things are any better now?

Regarding its own work in the regions, the EA has teams that go out in response to emergencies or reports of pollution. It is responsible for ground and river water, has a fisheries science team that measures river quality and undertakes biological surveys where ecologists assess invertebrate life in rivers. It also checks on flood defences and enforces waste management regulations.

Where foot-and-mouth disease is concerned, it is the Agency's responsibility to direct where carcasses can be buried, and where the waste from cleansing and disinfecting can be disposed of. For example, DEFRA will contact them to ask whether they can dispose of disinfectant etc. by spreading it on a particular piece of land and the EA will look at the map to see if water on that land will disperse slowly enough for the materials to biodegrade before they enter the water courses; then give, or refuse, permission.

According to David Edwardson, a regional team leader in the north interviewed by Jane Mayes on 13 July, most of the disinfectants used are citric acid based, so there have been no detectable problems in drinking water, for fish or for invertebrates or algae. However, he did admit that he was 'obviously aware that lots of teams are going out doing the cleansing and it can't be guaranteed that the proper processes are observed all the time'. Also 'that there is always the potential for one particular problem or accident to have long-term effects'.

What stories there are concerning water pollution are almost entirely anecdotal: that several sewage works were knocked out following pollution from disinfectant; a thousand litres of disinfectant that leaked from a farm or abattoir into the River Ewer in North Yorkshire; that there were dead fish in the River Dee near Chester but there are no reports of any such incidents on the old MAFF website.

Then there is the possibility, however remote, of water pollution from burial sites. On 24 May Professor Peter Smith, chairman of the government's Spongiform Encephalopathy Advisory Committee, which studies the effects of BSE, admitted publicly that he would not drink tap water near any foot-and-mouth burial sites. He warned that there was a risk, albeit a small one, of developing the human form of the disease, CJD, by drinking contaminated water.

In a statement quoted in the *Daily Express* and other papers on 25 May, he estimated that the risk of developing CJD from contaminated water could be as high as one in 200,000 in burial pits where older cattle had not first been incinerated. In other sites, where cattle have been burned and then buried, the risk is around one in a million. 'If there were a large number of animals buried over the age of five years I would not be happy drinking the tap water nearby.' He stressed that the risk of contamination is only from cattle over the age of five years – 'but it is still very real. There may be some contamination of the water, which would have its most immediate impact on people living nearby. If animals are burned on a pyre and their ash is buried, we assessed the risk to people living nearby. It would be one in a million, which is a small risk but it is still a risk. But it would be six times that if the carcasses were not burned. Next to a burial pit wouldn't be the first place I would want to live,' Professor Smith told reporters after meeting with officials at the Department of Health.

His warning came on the same day that it was revealed that CJD had claimed its 100th victim. Only six victims are still alive. It was also reported that the 98th death had been that of a nephew of the former Liberal Democrat leader, Sir Paddy Ashdown. Needless to say people living near to such burial sites were appalled at Professor Smith's announcement, not least those living close to one of the biggest burial sites, that at Widdrington in Northumberland where some 128,000 animals had already been disposed of.

A local resident, Peter Kull, commented: 'Just when you didn't think it could get any worse, it suddenly does. We're wondering if the shadow of this plague will ever be lifted from our countryside. We have no confidence in the authorities . . . we have to live here, and our children of June it was revealed that three soldiers had become ... victims of the foot-and-mouth epidemic. They had all ... rare disease known simply as Q-fever. It is thought they ... en working on the disposal of carcasses. GPs and health ... in other parts of the country were circulated and told to ... for possible further cases of the disease and to monitor any ... admitted who displayed symptoms of an acute, 'flu-like pneumonia and to ask them if they had either worked on farms where animals had been culled or had taken part in the slaughter programme.

If so, then it was advised that they should be blood-tested as the disease can be fatal but presents with very few specific symptoms. The three soldiers had all been involved in the slaughter programme in Northumberland. Two of them who had worked together on several farms in the area were admitted to hospital with the symptoms and breathing problems on 6 May and 11 May. The third was still under treatment at the time the story was publicised.

According to a note by *The Guardian's* health correspondent, James Meikle, on 28 June, Q fever is caused by an organism, Coxiella Burnettii, usually through direct contact with animals – sheep, cattle and goats. Placenta, birth fluids, blood and faeces are the most infectious substances, but raw milk has also been implicated. The fever is also spread through air or dust and the organism can survive for months. It is, however, relatively rare, though this might be due to under-reporting. Sufferers may present some, or none, of the symptoms: headache, fever, muscle pain and sometimes pneumonia. Complications can develop, a potentially fatal inflammation of the heart in 11 per cent of cases, or liver problems in 1 per cent of cases.

The three cases were not reported by the Department of Health. Details appeared on the website of the public health laboratory service which monitors infectious diseases. When this was queried, officials at the Department of Health declared that the general public were unlikely to be at risk of contracting the disease from the burial of animal carcasses. This was in spite of the fact that its own risk assessment, prepared earlier in the crisis, suggested that bacteria might be released from burial sites and also that the local populations could be at a low risk of exposure through contaminated dust and food.

When a spokesman from the Department of Health was finally tracked down and asked about the disease, he replied that it was known that the people who had developed it so far had not contracted it through food and that they were directly involved in the disposal of dead animals. 'It is a known risk. The disease itself is not a high-risk disease. The exposure among people who work on farms is very high because the bacteria are so common and widespread.' The public health laboratory has also warned doctors to look out for psittacosis, which can be spread by sheep and is known to be a risk to pregnant women.

So, from water to fire. Just how concerned should those people be who had to live, sometimes for several weeks, surrounded by thick, unpleasant smoke? For the first few weeks that the pyres were burning, the authorities – local and national – assured the general public that there was nothing to worry about, no hazard to health. Confidence in these pronouncements, however, did not last long (if indeed it ever existed), when it was revealed after the public meeting over the Hallburn airfield pyre at Longtown that virtually no monitoring had been done anywhere to discover just exactly what was in the smoke.

It was left to independent researchers to look into that. James Bell, the scientist who perplexed the panel at the Longtown pyre protest meeting, has provided the most comprehensive information in his paper 'Air Pollution from FMD Pyres'. The pyres were largely composed of quantities of combustible material, including railways sleepers, oil and coal, forming a bed for the carcasses piled on top of them. Typical temperatures reached in such pyres range from 800 degrees centigrade at the bottom to 1900–2600 degrees centigrade for the flames above the bed. Flames develop from the combustion of Volatile Organic Compounds (VOCs), particularly hydrocarbons, and the temperatures depend on the particular compounds involved.

A conventional foot-and mouth disease pit-burn is a relatively low temperature combustion which reduces the production of toxic emissions and therefore measurements taken of emissions from such sources will not show accurate data to assess high temperature incineration (i.e. over 500 degrees centigrade), which has the potential to create organic toxins. Among these are Polycyclic Aromatic Hydrocarbons (PcAHs) which are formed during combustion of the aromatic hydrocarbons in the coal and bituminous sleepers used as fuel for the incineration. Most of these are proscribed under both EU law (R45 substances) and United States Environmental Protection Agency (USEPA) regulations as carcinogenic.

Further, blood and disinfectant both contain chlorine which forms several dangerous toxins when incinerated, not least of which is Dioxin, which is a known and proscribed teratogen. It was the dioxin contaminants in the 245T and 24D, rather than the chemicals themselves, which made up Agent Orange, that are now blamed for consequent soft tissue sarcomas and birth defects. Dioxin was also fingered as the possible cause of birth defects some years ago in an area around a chemical plant at Bonnybridge that was later closed down. Under the limited oxygen control conditions on the pyres, carbon monoxide will be formed and also possibly phosgene (mustard gas). The sulphur content of the coal might well also produce sulphur dioxide and Furan, both of which are toxic to humans. Lastly there is also a large emission of VOC which will contain aldehydes of which formaldehyde is carcinogenic: this is already known from rendering plant emission data.

No environmental assessment, which is a requirement under UK civil law, was carried out on the proposed use of pyres to dispose of carcasses and no State of Emergency regulations were in force to render such a requirement unnecessary, The only licence sought for the Hallburn airfield site concerned groundwater, not air pollution.

It was the questions raised by James Bell at the Longtown meeting which caused such a furore: what is the result of the site Environmental Assessment specifically in respect of the release of VOC, proscribed PcAH, Dioxin, Furan, Smoke Particulates, Carbon Monoxide and Carbon Dioxide? What Civil Permissions had been granted and under which procedures? Was there any published data from previous similar burns to formulate controlled temperature range profiling, control of emission and fuel contaminant effects? Had the potential fallout area been assessed? What meteorological data had been subjected to independent appraisal and confirmation? What temperature monitoring and control systems were in use at the plant? What emission controls were in use? What remedial systems were in use to prevent groundwater pollution or draught-aided airborne dispersal of disease? Finally, the supposedly expert panel was asked if they understood the long-term effects of teratogenic and carcinogenic substance release into a populated area.

It was then pointed out that there should be sufficient expertise among those present to answer these questions in full as they were simple enough to answer if they had done their jobs properly. No answers were forthcoming. That was in April. By July there had still been no answers.

At the end of April an independent research organisation, Westlakes Scientific Consulting, reported its findings in the area of the Lake District covered by Allerdale Council, an area with one of the highest concentrations of outbreaks and consequently one of the largest number of pyres. The research showed that on a number of occasions short-term levels of potentially deadly chemicals exceeded national air quality standards. While the study concluded that average concentrations were within accepted levels – a fact hailed by some authorities as proof that there was therefore no hazard – the researchers said that a big question-mark still remained over the wider atmospheric impact and that a nationwide survey was needed to investigate more distant airborne pollution from the pyres. Westlakes Scientific Consulting is British Nuclear Fuels, now independent, laboratory which has moved off site but is still funded and staffed by BNFL employees, which led to one enquirer asking whether or not there should be a need for caution in burning sheep from the Duddon area, which are still under the 1986 Chernobyl restriction. He was assured there was not. As an interesting aside, Major Alex Birtwhistle, who played such a high-profile role in the army operation in Cumbria, has now joined Westlakes in an honorary capacity.

While most pyres are no longer burning, questions are now being asked about the safety, or otherwise, of the ash. On 12 July, Buckinghamshire Council erupted in fury after it was revealed that contaminated ash from the north-east was to be taken to the county to be buried there. Buckinghamshire had gone to extraordinary lengths to keep the disease out and was much castigated for closing off its footpaths and bridleways although there were no cases in the area. So the proposal to dump more than 3,000 tonnes of contaminated ash from the notorious Widdrington pyre into a landfill site near the village of Calvert, nine miles outside Aylesbury, caused public outrage. Not least because it is known that Widdrington was used for the disposal of cattle most at risk of having BSE.

On the night of 12 July the first trainload of 60 sealed containers of ash left the Freightliner depot at Teesside bound for the Calvert landfill site. Asked what he thought about this, deputy council leader Bill Chapple commented: 'We are shocked that the government has said Buckinghamshire must take this waste through the back door, since we have remained free of the disease, despite many surrounding areas being infected. Where is the logic in taking 17 trainloads of pyre ash from infected areas in the north-east and dumping it here, which has always been infection free?'

The Teesside railhead will also be taking ash from the large pyres at Catterick and other northern sites, for shipment to Calvert. Dr Mike Tas, National Director of Disposals for DEFRA, commented: 'Although we have always been confident that the transport of ash through Redcar and Cleveland presented no health risk, we recognise that local people have been concerned and are pleased that disposal plans have now been finalised. Our most recent analytical information on pyre ash is very reassuring.'

Not everyone, however, is reassured by this assessment. One expert on the subject is reasonably sanguine about the safety of the Buckinghamshire site as that particular site has a substrata of clay (it is in the major extraction region for brick-clay), which would mean the site is naturally lined. He points out, however, that the removal and transportation will put the fly ash into the atmosphere as particulate matter — you can't avoid the dust raised by digging it up and moving it. Or tipping it into the pit.

The final question to do with the safety of pyre smoke is whether or not it might actually spread the virus, especially before the burn reaches a sufficiently high temperature, the question raised by the *Telegraph* correspondent who wanted to know if the infection had spread downwind from existing pyres. That, too, remains unanswered.

None of the above is good enough. People no longer believe the glib assurances that everything will be all right; that whatever contaminants there may be, they are present in such small quantities as to be harmless; that you would have to drink litres of contaminated water and stand in smoke for days to suffer the slightest ill effect.

People recall what was said about asbestos and the special form of cancer it causes in the lining of the lung – mesethelioma; a whole host of pesticides (like DDT) which are now proscribed; the years of assurances about BSE, how it could not possibly cross the species barrier and affect humans; how the scientists who blew whistles or gave warnings were excoriated for their pains. Few will forget the image of John Gummer feeding his infant daughter a beefburger to prove just how safe it was. Not to mention all those weeks when we were confidently assured that the foot-and-mouth outbreak 'was under control'.

It may well be that everything is all right, that there will in future years be no ill effects, let alone cancers or birth defects caused by contaminants in pyre smoke; that there is no trace of anything harmful in the groundwater and there is absolute proof that pyre smoke does not, in itself, spread the virus. Unfortunately, however, if there are serious problems it is unlikely that they will show up in the immediate future. If they do, it will then be argued that it cannot be proved that contamination is the cause of whatever has gone wrong; and anyway, no one had any reason to think there might be any problems. Or rather they hoped and prayed that there would not be any while they might still be held responsible. This philosophy is known as 'NIMTO' – not in my term of office. This is simply not good enough. At the very least the public has a right to demand that research and monitoring is undertaken as a matter of urgency should the present reasoning later be proved wrong.

# Chapter 13

# Aftermath – The Human Cost

The Porter in Shakespeare's *Macbeth* always raises a laugh when he expands to the audience on who it might be out there hammering on the door to be let in. 'If a man were porter of hell-gate, he should have old turning of the key,' he says. A stage direction (rare for the theatre of that time) informs us that the knocking continues.

'Knock, knock, knock! Who's there in the name of Beelzebub? Here's a farmer that hanged himself on the expectation of plenty.' Presumably even in Shakespeare's unsubsidised times he was familiar with local farmers who moaned and groaned even when the harvest had been good, and there truly was the expectation of plenty. The portrait of the whingeing farmer, not always unjustified, has therefore passed on down the centuries. Nor have all those interviewed on television or radio met with a great deal of sympathy from the urban population. A case in point, and one which caused much general discussion on the subject, was footage in a BBC newscast of a farmer who had suffered a cull and who was shown outside his splendid, picturesque farmhouse with a new Range Rover or similar vehicle parked outside. 'I bet he suffered all the way to the bank when he took in his compensation cheque,' was a common reaction.

But this is certainly not the experience of many, some of whom have quite literally been driven to hanging themselves. It is impossible to know what the long-term aftermath of the worst foot-and-mouth epidemic in the world is likely to be. Although it is clearly too early to hope that we have even reached the aftermath. But in the short term, the consequences are far-reaching and go well beyond the obvious financial loss. Anyone who has spoken to or interviewed farmers, especially those suffering from restrictions or having recently experienced the slaughter of their livestock, knows just how desperate they can be. Desperate to talk, to tell the story over and over. Below are two viewpoints, from an extremely wide range.

The Reverend David Mewis, the Rural Dean of 21 parishes in Cumbria, speaks movingly of the effect it has had on his parishioners. 'Our community has been in isolation. It has been a massive bereavement where people have been on their own. Normally, if someone dies, people can surround them. I can go and see them face-to-face. With foot-and-mouth, you're locked in, isolated, the people round about you can't get to you. You can't even send your children away while your animals are slaughtered and often these include animals which have become the children's pets. You can't meet up with other people, or get well away from it all.'

Those outside the immediately restricted areas are isolated too. 'They go into the towns or villages shopping just so they can meet up with someone else.' Not surprisingly people have turned back to their local church. 'Our church numbers have gone up. People coming out of church at 12.15 p.m. are still there talking at two o'clock. There is a desperate need to talk, to share experiences.'

Mewis is particularly concerned at the effect which seeing their animals slaughtered has had on so many of his parishioners. He recognises that in some cases this was managed with as much sympathy as possible, but 'many slaughter teams simply did not have any compassion for the animals. In fact a lot don't. Their view is "we've bought them, they're ours, it's up to us what we do with them," and then they try and do things like attempt to move a herd up a stream bed for instance. A farmer who knows his animals wants them put down in an orderly and compassionate way. He then sees what the culling squads do and they tell him to keep quiet.' Mewis thinks their whole attitude is wrong. 'They shouldn't just come in, in a haphazard manner. It needs to be done quickly but compassionately, even down to how you move the carcasses after the animals have been killed.'

The last sheep left alive in his parish are his own, just 11 of them, in a croft at the side of his house. 'If mine finally have to go, I want them treated with compassion, then buried. I want it done properly. We know it's a job that has to be done. But those who have put their lives into breeding animals for years and years, who show them with pride, have now seen it coming to an end in brutality. To see them slaughtered, not in a careful way but with brutality.' Often no notice is taken of the farmer's expertise, let alone his feelings, and this is a mistake. 'He knows how to deal with it, the proper way of doing it. You don't slaughter a calf in the sight of a cow. It has to be done decently.'

He worries too over the effect the experience and the images will have on the children who have had to see what has gone on, lived beside the rotting carcasses for days, then quite possibly seen them burned on their own land. Children who were marooned in their homes, unable to go to school, attempting to study in an atmosphere of utter desolation, then expected to go off and take major examinations.

He recognises that farmers will, of course, receive compensation for their slaughtered livestock, but for a long time they still cannot put any livestock out in the fields. Nor is it simply the farmers whose immediate livelihood has disappeared. 'There are the sheep shearers and the firms that hire out and sell portable sheep-shearing goods. They have nothing to do as there is nothing left to shear.' One of the worst aspects in Cumbria is that 'we have all had to stand by and watch the rare bloodlines go. One of my parishioners, a widow, had a bloodline which had come down three or four generations at least and her flock has now been culled out. You can't replace that kind of thing. Even when they restock, it will take generations to return sheep to the fells in that area. It doesn't just come by instinct, it comes by the passing on of knowledge.'

The aftermath, he says, will be like coming out of a bereavement, with little to look forward to. 'What can the small hill farmers look to now? "Rationalisation" of local farms? Some people will move away and with them will go the village shop, the pub, the school and everything else. That is what happens. The land looks tatty at the moment. It takes time and energy to maintain the pasture. It doesn't just stay like that by chance. If it isn't maintained it goes back to nettle and dock, not wild flowers. That's not the nature of this land . . . The high fells are limestone and without grazing will turn back into scrub fairly quickly. What the tourists and walkers want to see is actually a cultivated landscape with animals, especially the rare breeds.' But even if the epidemic really has reached its end (and he, like many others, is very doubtful), there will be nothing to see even next spring. 'By the time it's possible to start restocking with sheep it will be November and that's too late. There'll be no lambing around here next year, it's too late to take the sheep to the ram.' He expects it to be at least two years before the livestock are really replaced, and in the meantime what are people to do? He is suspicious too about the sheer numbers of livestock slaughtered and how his own knowledge conflicts with published information. 'Yes, so in one part of my area we have had had four published cases, but in fact we've lost 17 farms. And suddenly all the numbers have gone, edited out, so we don't realise the full impact. Yet I've seen the livelihood of 17 families disappear overnight – that is the kind of impact foot-and-mouth has had on this area.'

Farmers had already been pressed to diversify, to become part of the tourist industry. 'Some spent a lot of money on cottages and so on for letting and bed-and-breakfast, etc. and now they are empty. The only people using them are from MAFF or now, DEFRA.'

One of his parish priests was, eventually, licensed to go out and offer support. 'For people still die or give birth within all this.' He has 54 farming families in his two parishes alone, apart from those in the 21 deanery parishes. 'We now have to help bring them out of their bereavement and it's going to be a long job.'

Sheep farmer Lisa Robertson, writing in the *Western Morning News* in mid-July, reinforces that feeling of isolation and lack of hope. Already, before the outbreak, she had watched the price of the stock she had reared go down and down as the buyers for the supermarkets controlled the bidding. Now she no longer has any animals to put up for auction since her entire flock was slaughtered back in May.

When she was told that her sheep had tested positive for antibodies, she rang a friend, a leading virologist, whose response was 'rubbish!' Her sheep had been kept indoors since the epidemic began and had lambed successfully. She had brought in no new stock for two years. 'They've bungled the tests,' her friend told her, 'I'll phone London and ask for a retest.' A quarter of an hour later he rang her back, very upset. The spokesman at MAFF headquarters had shouted at him, 'They must die, they must die!'

She found it hard to understand how her sheep could have contracted the disease when none of the dairy farms around had gone down with it, but this was to no avail. Everything had to go, even her pet goats. She asked a MAFF vet if she could have a retest of one of the sheep and have it checked out at Pirbright and returned quickly – if Pirbright confirmed the diagnosis, then she would go along with it. He agreed, as he could not believe the flock were infected. She heard him tell MAFF in London that he thought they were making a mistake, 'a political mistake if nothing else'. London refused to budge. They would not allow a retest.

Isolated as she had been on her supposedly infected farm, the memories will haunt her for years. 'My breeding ewes were my friends, they gave me the most beautiful lambs year after year. I knew each one individually and had nursed them through cold wet nights with triplets, difficult births, even pneumonia and they all lived.' She had gone out to round them up with her sheepdog, knowing her leaping flock would be dead within the hour. The only good that came out of it, she feels, is that she and Noel Edmunds fought what finally became a successful campaign for a proper blood testing policy, including retests.

✳ ✳ ✳

For some, life became so intolerable that they have taken their own lives. This has happened in several parts of the country but in Powys alone, the county which has borne the brunt of the epidemic in Wales, three farmers committed suicide between the end of March and the third week in April. Glyn Lewis from Llwyn-y-maes hanged himself in a shed full of cattle on 21 April. Coroner John Hollins told the inquest, one of three held in Welshpool town hall, that Mr Lewis had been so careful of spreading infection that he had not used his lorry for seven weeks because he was afraid of carrying the disease to unaffected areas. However, despite his own livestock proving healthy, he had just learned that they were about to be culled when he took his own life.

A local police constable, Matthew Thomas, told the inquest that Mr Lewis had become worried when he could not help others. 'He was suffering because of the constraints of foot-and-mouth. He did not want to take the risk of bringing it to the area. His cattle were to be culled in the next couple of days and his business was a huge part of his life. He was a very, very popular man and being the sort of person he was, he wanted to help everyone then worried when he could not.'

The second farmer was John Bayliss, found dead from a single gunshot wound to his head at his farm near Newton in April. Although none of his cattle were carrying the disease, the farm experienced the usual restrictions because a contractor who had been on an affected farm had visited it. Because of this he was unable to move his stock and his situation became similar to that in north Devon during the same period, with sheep and lambs dirty, wet, smothered in mud and unable to be moved. He had dwelt on this. 'There were lambs in this state everywhere,' his wife told the inquest. He had simply been unable to take any more.

The third farmer, Brian Oakley, already had a history of depression. He had suffered a breakdown in the early 1990s caused by the slump in the industry and had never properly recovered. They had been forced to move to a smaller farm. His mood, his wife told the inquest, had definitely been affected by the epidemic and he had gone out and hanged himself.

The coroner called on the farming unions to find 'some way of reaching out and supporting members of the farming community affected by the recent crisis . . .The crisis in the farming industry has been catastrophic for the families of these three decent men and their deaths are to be deeply regretted.'

We will look in the next chapter at the economic implications of what has happened, but what is so striking when one talks to the small farmers is that after weeks, possibly months, of anxiety, isolation or semi-isolation, they can see no real future ahead. Lisa Robertson says that the implications of what will happen this coming autumn and winter are dire. 'If the smaller farmers cannot scrape a living, they will be forced out of the industry and their experience will go with them; for ever.' The government is asking us to steward the land, she continues, and reduce the numbers of stock. But if the farmers go, who will be left to steward the land?

There seems to have been very little concern about the mental health of those affected by the epidemic. Like everything else this is a problem which is ticking away for the future. With no livestock on farms for months, and little opportunity to breed during the autumn, 2002 will usher in a silent spring. As the Reverend David Mewis puts it: 'The countryside has become a great emptiness.'

# Chapter 14

# Aftermath – What Future for the Countryside?

Whatever happens and however long it takes, the epidemic and its aftermath will usher in change whether wanted or not. What kind of change, however, is likely to be a matter of extreme contention with various groups vying against each other and a government which is perceived, maybe unfairly, always to be swayed by the biggest, most powerful and wealthiest lobby. We are likely to see more of town versus country, farming interests versus the tourist industry, big farmers versus small ones and always with the suggestion that there are other agendas we do not know about.

Let us put all the special interest groups aside for a short while and consider the likely attitude of the average taxpayer, living in an urban environment, struggling to pay a mortgage on an overpriced house and taking little interest in where his or her food comes from, as long as the local supermarket is doing a good offer this week. He or she is unlikely to be sympathetic to large sums of money being paid to rescue a rural economy brought to its knees by what is generally agreed to have been gross incompetence.

For a substantial proportion of the population do consider farmers to have been feather-bedded for too long. Again and again one hears it, even when researching the foot-and-mouth epidemic. 'They didn't rescue the miners, did they, even if one day we'll find ourselves short of fuel?' 'What about the steel industry in Wales? Throwing more men out of work in a region where there are still pockets of high unemployment.' 'I'm self-employed, no one's going to pick up the pieces for me if I lose all my work.' And make no mistake, large numbers of townsfolk do feel this way.

They view bodies like the Countryside Alliance as being made up entirely of well-heeled farmers who have consistently milked the system for years and who leave the hard work to a handful of workers while they ride to hounds every other day. They may have flinched at the appalling images of twisted corpses on television, perhaps even gone off meat for a few weeks as a result, but have remained basically unsympathetic to farming in general even though some of them rush off into the countryside to buy second homes, raising property prices and making it even more difficult for local people to stay on the land.

Taxpayers will have to pay out at least £2.5 billion for government spending on foot-and-mouth, according to a report in *The Independent* on 1 July which has not been officially denied. (This could be a major underestimate, other sources now put the final bill at four times that.) That is the equivalent of putting one pence on the basic rate of income tax; an amount, the paper points out, that could have been used to dramatically cut hospital waiting lists and class sizes. Around £1 billion has been pledged to farmers in compensation for the mass slaughter and another £1.5 billion has been spent on financing the vast operation involved in killing and disposing of the carcasses. This does not take into account the enormous loss of revenue to the rural economy in general and the tourist trade.

This figure completely dwarfs the value of the export trade – roughly £570 million – that the slaughter policy and the contiguous cull were designed to protect. The money spent on that alone would have financed 10,000 hospital beds, wiping out the shortage, as Liberal Democrat MP Matthew Taylor, his party's Treasury spokesman, points out. 'The government says it has to resort to the private sector to fund schools and hospitals, yet it has been able find, without blinking, the equivalent of a penny on income tax to fund the mishandled foot-and-mouth crisis. This surely proves the case that a full independent inquiry is needed into its handling of the foot-and-mouth epidemic.'

Asked for his comment, Charles Secrett, Executive Director of Friends of the Earth added: 'At last we know the real cost of the incompetence of the old MAFF and of Tony Blair's blind allegiance to agribusiness.' On the same day these figures were announced Holland, the second hardest hit country after importing our infected sheep, quietly resumed its meat exports. It had wiped the disease out. It is against this background that the various interest groups are now making their pitch.

* * *

The debate about vaccination is taking place in an atmosphere of almost total distrust and in farming quarters, most particularly among small farmers, there is constant talk of a hidden agenda.

Rumours abound: that trials had been running in four countries prior to the outbreak, one of which was the UK, which involved injecting live foot-and-mouth disease virus into pigs, thus infecting them, and that the virus then 'escaped'; that the disease was started deliberately to cut down the number of livestock in the country and that even if that was not the case, then it was allowed to let rip for the same reason as this suited proposed government policy; that back in 1998, Germany and the EEC had demanded we reduce our livestock by 30 per cent and go over to more arable farming, a policy greeted with enthusiasm by the agribarons and the executive of the NFU. That even if the disease had come out of the blue and everything possible had been done to halt it (generally agreed to be incredible), the Blair government will use it as the excuse it has long been waiting for to shaft the small farmers and put them out of business.

Asked back in June about the possibility of a hidden plot to put small farmers out of business, Anthony Gibson responded that it was complete rubbish. 'There is no logic to it because the EEC, government, etc, don't support farmers, they support production. If you get rid of small farmers you don't save any money, as the nearest big farmer will take over his land and might well produce even more. It wouldn't solve any problems at all. The subsidy is per hectare of cereals or number of livestock, not the farmer.' There would be no benefit, he continued, in getting rid of a third of all farmers as, if people lose their jobs, they immediately become a charge on the exchequer. The notion of farmers going bust is, in fact, a disincentive.

However, even he was somewhat alarmed when Lord Haskins, described as 'the Prime Minister's key agricultural adviser', declared that what Britain needed was ever bigger farms and to this end subsidies should be scrapped all together. Lord Haskins, whose expertise in this field is not revealed, is chairman of Northern Foods (which hardly makes him an unbiased observer) and has been made head of yet another government task force, the 'Better Regulation Task Force'. Farmers, he declared robustly, have been 'mollycoddled' for too long.

His lordship's comments were made in an interview with the *Daily Telegraph*, during which he said that he would expect there to be only half the number of farmers working the land within the next 20 years. 'Farms will get bigger and that's a good thing. A lot of agricultural reformers, like the Prince of Wales, want farmers to stand around being subsidised and making thatched roofs. Well, that's for the birds. Agriculture has got to strive to be more competitive and more productive.' Lord Haskins has promised us a map of Europe, showing where farming can be considered competitive and where some agricultural communities might need extra support.

So, what he appears to be proposing is more of the very same developments that many believe have led to the current crisis: even bigger prairie farms to enable the agribarons to produce even more unwanted cereals; larger concentrations of livestock which will continue to be moved all round the countryside, with all that implies in terms of risk; while the smaller farms and a good half of the organic side of the industry goes to the wall with or without thatched roofs and birds.

In response to Haskins' interview Anthony Gibson replied that the peer's comments were 'completely misguided. I do not know anyone who thinks that the future of British farming lies in cow factories and prairie farming. Much more alarming is that he is one of Tony Blair's key agricultural advisers. If these are the views the Prime Minister is being given, then it is very worrying indeed.' It would, he continued, devastate both the environment and the wildlife. 'The very thing that makes up the fabric of the countryside would be lost.'

Farmer Richard Mawdsley, commenting from devastated Cumbria on Haskins 'that's for the birds' remark agreed that yes, it was for the birds. 'For the birds that have so disastrously disappeared from our countryside. It's also for the lambs, cattle, pigs and goats that are in the middle of a still unpublicised holocaust. It's for the traditional farming that has shaped our much-loved landscape. It's for food that tastes wholesome rather than mass-produced pap for the masses. And it is for Britain, rather than Europe's expansionist ambitions.'

The leader writer in the *Western Morning News* also put it succinctly: 'If Lord Haskins thinks all farmers are mollycoddled, he should take a look at the state of agriculture in much of the West Country, where the smaller, hard-pressed family farmers were suffering even before the foot-and-mouth outbreak. He should take a trip to a town like Hatherleigh, devastated by crisis after crisis in the all-important agricultural industry that supports that community's entire economy . . . And he should talk to some of the dedicated livestock farmers who escaped foot-and-mouth but were left with virtually no income and no compensation as the disease forced markets to close and shut off virtually all means of earning a living. If that's mollycoddling then you can keep it, they would tell him. And not in such polite language either.'

Having said all this, no one doubts that there is need for radical change and when the disease was raging at its height it did look as if some consideration might be given to the overall state of the industry, whether it was sensible or necessary to move livestock around in such numbers or if it would not be better to ensure that in the future most animals should be both kept, and slaughtered closer to home. But with the disappearance of the epidemic from the headlines, it is feared that the possibility of improvements to the system has gone with them.

As well as the rumours of devious plots to decimate the farming industry, there have, of course, been others from an opposite viewpoint: that hill farmers had been used for years to 'borrowing' sheep from each other when it came to counting heads for subsidy; 'The subsidy man's coming tomorrow, so lend me 200 sheep, then I'll do the same for you and Fred down the road'; that valuers so overwhelmed by the sheer numbers involved that they accepted a farmer's overestimate of the number of animals slaughtered since they simply did not have the time to count them; that slaughter came as a relief to farmers already on the breadline as the compensation packages prevented them from going out of business just in the nick of time.

But with or without reform of the Common Agricultural Policy, a subject which now seems to have been grinding forever, change will come. It is the nature of that change that was already causing concern even before Lord Haskins made his insensitive remarks.

There is no doubt that DEFRA is regarded with a great deal of suspicion, not to say cynicism. The name may have changed and the Minister may have changed but some things at least remain the same. The senior civil servant responsible for handling the foot-and-mouth epidemic, with all that implies, has been substantially rewarded by Tony Blair in spite of the torrent of criticism levelled at the old MAFF. Brian Bender, who was Permanent Secretary to Nick Brown, has been given an even bigger empire in the new department. It includes all of DEFRA, the responsibility for the environment coming to him from the previous Department of the Environment, Transport and the Regions, plus animal welfare and hunting from the Home Office. As *The Times* pointed out when the post was announced, this one man is now in charge of foxhunting, climate change, genetically modified foods and animal rights . . .

He had only been at MAFF for nine months when it became DEFRA and is seen as a moderniser. 'Some are still critical of his performance over the past months,' says the report, 'while others question how well he will get on with his new Minister, Margaret Beckett.' To which one can only marvel at the choice of the word 'some' to describe hundreds, probably thousands, of critics. However, part of his brief is to move on from protecting the farming industry to protecting consumers. We shall see.

Margaret Beckett has, as they say, hit the ground running. Within hours of taking up her new post she had upset protestors in the West Country by declaring that it still might be necessary to use the vast burial pit at Ash Moor, in spite of Nick Brown, before he left office, promising that this would not happen. Presumably – though this was not made explicit – Mrs Beckett is saying that Ash Moor has to remain an option if the disease not only refuses to go away but starts spreading again.

Then, within a month, she had issued a warning to farmers that the traditional subsidies are no longer sustainable and would be phased out within the foreseeable future, choosing to announce this at a somewhat odd event: a national conference on the future of farming organised by the Royal Society for the Protection of Birds. Few would disagree with her when she said that the Common Agricultural Policy, forged 40 years ago, was now failing farmers, consumers and taxpayers. 'The CAP policies of market support and direct payment to farmers are very blunt, inefficient and ineffective instruments. Market price support and production controls are outdated mechanisms that should be phased out, with farmers being helped to adjust with transitional support.'

The appalling problems caused by the foot-and-mouth epidemic, she continued, had highlighted the problems of rural communities and accelerated the need for change.

Mrs Beckett was supported at the conference by the German agriculture minister, Renate Kunast, who said there was now a growing consensus across Europe that CAP would have to change. Ms Kunast, a member of the Green Party and a supporter of organic farming, said there might well be a case for subsiding organic and small family farms to achieve wider social objectives. France has always been the stumbling block to any reform of CAP but it was hoped that joint pressure from Britain and Germany might finally persuade France of the need for reform.

Many of her listeners, however, remained unimpressed. Jenny Blashford, organiser of a scheme for farmers in South Somerset, told reporters that it was virtually impossible for small farmers to take part in the government's agri-environment and organic schemes because of the costs involved. She told of one farmer she knew who considered converting his 100-acre farm to organic production but had to abandon the idea because it would have cost him around £100,000, which was unsustainable for a small business. 'I am working at the coal face,' she said, 'with people facing very difficult circumstances at the moment. There is really nothing in all this for small farmers. We need positive discrimination for small farmers – they are missing out on a huge amount.'

Supporting her, Philip Hosking of the Small Farms Association, said many small farms were already carrying out the environmental measures the government was seeking, with no payment or recognition for their efforts. 'If farmers were not looking after the environment you would soon know about it. At the moment we are missing out.'

Anthony Gibson was even blunter, describing the speech as a 'wish list embroidered with buzz words, not so much spin as knitting'! Farmers recognised the need for change, but they wanted far more detail as to just what was being proposed. The government must recognise that the impoverished state of the industry meant many farmers could not afford to change. Indeed without short-term aid, many might not even survive long enough to bring about the changes the government wanted. The situation was now so serious that the government had to realise it could jeopardise all their long-term aspirations.

He was speaking for farmers in the south-west, but Mrs Beckett's words went down like a lead balloon in Cumbria too, where news had just emerged that the government would find themselves with between 1.5 and 1.7 million 'surplus' lambs by the autumn. The hill farmers still left with the sheep wonder how this can be, since the national flock has been reduced by 2.8 million, 15 per cent of the total, since December 2000. It is estimated that the amount of lamb sold to domestic customers is 194,000 tonnes, a further 115,000 tonnes being imported from other countries. So where is the surplus?

A spokesman for breeders in Cumbria said:

> For any government to consider the purchase and destruction of these lambs is not just ridiculous, it verges on the criminal. It is on a par with a greengrocer who owns a potato patch behind his shop, but who buys from a wholesaler because he's too idle to dig his own. It is the government who, in their ignorance and passion for bureaucracy, has made it impossible for us to conduct our

business . . . In the wake of CAP, BSE and FMD the British rural economy, based on Agriculture and Fisheries, is in a state of collapse . . . we were asked to produce more, we did. We were asked to produce even more again and in our naïvety we did, despite decreasing margins. Then, after years of requests and inducement to increase production, we get told to produce less. And to add insult to injury, to watch as food is imported, produced cheaply by methods that would render us liable to prosecution! Despite legislation being in place to prevent this happening.

\*  \*  \*

But as we know, while the farming community may have borne the brunt of the immediate horror caused by the disease, the ramifications for the general rural economy go on and on. Media attention has focused on the hotels and boarding houses, but that is only the tip of the iceberg. Whole swathes of activities had to be cancelled, from agricultural shows to a yearly ballooning extravaganza in Cumbria, leading to its organisers going out of business. Riding schools lost almost all their customers, as did those providing accommodation and/or equipment for walkers. And in spite of all the current spin to the effect that holidaymakers are flocking back and things are almost back to normal, anyone who lives in or near to a tourist resort knows that is simply not the case. Business in pubs and cafés in many areas is substantially down on what might be expected at this time of year, vacancy signs hang in the windows of bed-and-breakfast establishments and renters of self-catering properties have simply not turned up.

A large proportion of footpaths have now been opened up but many of those closed for months are now so overgrown as to be almost impenetrable and walkers are still regarded with suspicion by much of the farming community. In Dumfries and Galloway, the Lake District, Devon, Cornwall and mid-Wales, businesses are now facing the prospect of reaching the end of the holiday season, still severely out of profit, and with nothing to look forward to except worry about how they will get through the coming winter.

Jeremy Roe, the barrister with holiday interests mentioned in a previous chapter, has now launched a new organisation, Action for Foot-and-Mouth, believing that the many businesses damaged by the epidemic have a far better chance of exerting influence to prompt remedial action if they act together, rather than trying to press for it separately. So far, he says, rural businesses have got nowhere in asking for compensation for the losses incurred through no fault of their own, and one of the organisation's activities will be to investigate the possibility of taking legal action for the damage they have sustained.

He puts the damage to the tourist industry and rural economy at £5 billion, twice the amount suggested by *The Independent*. The reason why this happened, he says, is simple. 'The government acted as it did to protect the farming industry against losses a fraction of the size of the losses caused to the tourist industry and broader rural economy. To protect the farming industry against the economic consequences of a non-fatal animal disease, the government caused many billions of pounds worth of damage to other parts of the economy. Of course the government had to deal with the outbreak but there were other remedies available, remedies that have been used successfully by other governments. And given what has happened, who believes that the farmers have been helped by the government's actions?'

Most damaging to the government's case, he continues, is that a series of decisions were taken without the interests of the tourist industry and the broader rural economy being taken into account. 'In other words one of the key examples of negligence is that the decisions to cull over four million animals and close the countryside, were taken from the perspective of farming alone. The interests of the tourist industry and the broader rural economy were not taken into account.'

The government has rejected any responsibility for these losses and had deliberately promoted the following messages: that it did everything it could; that the scale of damage to the tourist industry and rural economy was unforeseen; that it was consequential; that it isn't as bad as initial estimates suggested and that all that is needed to remedy the damage is to spend a few millions on marketing.

To these assertions Roe responds that the answer to the first statement is 'nonsense'; to the second, that the scale of the damage was obvious, and was direct, immediate and foreseeable; that the damage is worse than first estimated and that substantial aid is needed, not just money for advertising. 'The government will talk to us and sympathise with us and commiserate with us and be photographed with us. But it will not voluntarily help us.'

Roe emphasises that this simply must not happen again and suggests that the government has to have a contingency plan for the future. To this end a Foot-and-Mouth Contingency Plan is being drawn up by the South-west Rural Task Force in the hope that it might be adopted as a national model so that if the virus does flare up again in the autumn, then such a plan could help minimise the economic and emotional damage. So far as farming is concerned, the plan examines current government policy and looks at various options with regard to vaccination and/or slaughter. But it goes beyond the single issue of the effects of foot-and-mouth on farming and into how it affects the broader rural economy as a whole.

Here, at least, vaccination is given serious consideration and several options are recommended for discussion, for example: creating a cordon sanitaire around outbreaks, and in order to be able to better control the disposal of carcasses; blanket vaccination across the board for use in place of mass culling, or as a permanent protection against slaughter for economically sensitive animals, such as those in zoos, though possibly rare breeds should also be considered in this category.

James Bell from Longtown, victorious against the Longtown pyre project, has also suggested a crisis plan for future use should it be necessary once again to slaughter animals and dispose of carcasses:

1. All farmers to declare annually their capacity for animal burial on their own land as a proportion of their stock, taking into account the regulations and ground water.
2. All waste disposal companies to declare annually their capacity to take infected stock.

3. All renderers and incineration facilities to declare annually their capacity to dispose of such animals.
4. All sheep to go for burial.
5. All bovine stock under 30 months to go for burial.
6. All bovine stock over 30 months and under five years to go for rendering.
7. All bovine stock over five years to go for incineration.

Which fits in neatly with the south-west region contingency plan where it is stated that 'under no circumstances would there be large-scale burning on pyres'. Both Bell and the South-west Rural Task Force also emphasise the need for efficient, open communication and better ways of disseminating information in the future.

While outside London much thought is being given to what needs to be done in the future there is no hint that the government is thinking that far ahead. As Jeremy Roe puts it, 'at the moment no one is convincing the government that they have an effective opponent. And the government will do nothing to help us unless it is forced to.'

His suggestion to concentrate minds is for the broader rural economy to band together and demonstrate that it possesses legal rights and both the resources and the will to enforce them. He believes that businesses damaged by the government's response to the epidemic can sue for damages on the grounds of negligence and nuisance. Also, that there are grounds for a Judicial Review regarding the legality of the cull of healthy animals possibly on the grounds that it is irrational, i.e. that it is irrational to take steps to avoid damage in one industry if they cause many times that amount of damage in others.

But all the above, the final cost to the taxpayer, the future of farming, the survival of the rural economy are currently based on two premises: that we know everything that is it is possible to know about the 2001 foot-and-mouth epidemic to avoid it happening again, and that it is really and truly over. Regarding the first, in spite of all the clamour, at the time of writing we are told there is to be no open public inquiry, and as to the second, it is clear from the recent rash of outbreaks in Northumberland that the nightmare is far from over.

# Chapter 15

# 'A Crackpot Idea' –
# The Need for a Public Inquiry

If ever there was a case for a full public inquiry into the causes of a major disaster, its handling and its aftermath, then the 2001 foot-and-mouth epidemic is surely it. It has cost the country yet unmeasured billions. It consumed the media for weeks in the early spring and postponed a general election. At every stage there has been heavy criticism of the way it has been handled and it has devastated great swathes of the countryside. As things stand there is absolutely no reason to believe that lessons have been learned. That it could not all happen again in almost exactly the same way, with the same lack of foresight, chaotic muddle, mismanagement, and mindset against considering other options as to how such an epidemic should be handled. To almost all observers there is only one way to prevent this and that is to ensure that every aspect of what went wrong is fully, and publicly, aired and that requires a public inquiry.

Yet within a week or two of taking over at DEFRA, Margaret Beckett was already stating that a full-scale public enquiry would not be necessary. What was needed was 'a different kind of inquiry', on the grounds that setting up a public one would take too long and distract from efforts to halt the epidemic. She was quite sure that what the farming community and rural community would want to see was an inquiry that was carried out fully, 'that is effective, but that deals as expeditiously as possible, and at as low a cost as possible, in uncovering the answers'. No mention here of public hearings or openness, nor on what basis she is so sure that this is what the rural community wants when all the evidence is to the contrary; though no doubt it is exactly what the government wants.

Next to wade in was Alun Michael, the MP who had been selected by Tony Blair to head the Welsh Assembly until his position became untenable and who is now Minister for Rural Affairs. Speaking in a debate in the House of Commons on 4 July he told the House that 'it would be irresponsible to bandy about words such as "public inquiry" which have specific meanings. The use of such terminology is not casual and the worst possible outcome would be to end up with a long-delayed and bureaucratic inquiry.'

Liberal Democrat MP Colin Breed said it was nonsense to suggest that all those calling for a public inquiry were behaving irresponsibly. Indeed Michael Meacher, in his capacity as chairman of the government's Rural Task Force, had said publicly on 3 April that it was clear that a full-scale public inquiry would be needed into such a major national crisis in order to ensure that all the lessons were learned. Breed pointed out that concern was growing over the government's attitude towards the pressure for a public inquiry, stoking fears that the government was ducking out of bringing the whole matter out into the open.

'It is not irresponsible at all to call for a public inquiry. The reason we want a public inquiry is because that is the only type of investigation that will sit in public and that will be able to call Ministers and former Ministers. Most of the former MAFF Ministers have now gone and unless the inquiry can call them it will be of little value. It is quite clear that the government are going to try and hold the line on this, which is that they will have their own sort of inquiry when the epidemic is finished. But many of us believe that they could get going on some things right now.' Pressure had started before the election and the change from MAFF to DEFRA.

The day before the general election two Devon women, Sarah

Mitchell-Jones and Sarah Waldron, took a 20,000-signature petition to MAFF signed by people who were opposed to the way the government had been handling the crisis, and calling for a public inquiry. The *Western Morning News* had asked people to phone in to say 'Yes' or 'No' to a public inquiry and the response was 2,242 for and 61 against.

Two bishops waded in. The Bishop of Truro, the Rt Reverend Bill Ind, and the Bishop of Bath and Wells, Jim Thompson, both came out in support of a public inquiry. The Bishop of Truro, who had spoken movingly of the effect of foot-and-mouth in a special service held in Truro Cathedral (at which the Prince of Wales also spoke), commented later that there had been such an inquiry after the 1967 outbreak which had made 16 recommendations, 'but they were totally ignored this time round'. He argued that the rural community, which had been so hard hit, deserved to be heard.

Bishop Jim Thompson has been a popular regular contributor to 'Thought for the Day' on the *Today* programme from the time he was a bishop in the East End of London before moving on to Bath and Wells. But on this occasion he appeared on the programme explaining why it was essential that a public inquiry be held into the whole catastrophe. 'I have rarely seen such devastation and breakdown of trust between the rural community and the government,' he told listeners. 'The cost of a public inquiry is as nothing set against the cost of the epidemic.' Such distrust was 'fouling up' the inevitable and important reconciliation between government and the rural community and a decision was now urgent. By that he did not mean that it should begin immediately but that it should be committed to immediately. The epidemic had been a major disaster for the whole of society. 'We need to know both the causation of the thing, and how it was dealt with, not to put blame on people but to discover for the future how it might be done better.'

Nor have all backbench Labour MPs sat meekly by their pagers waiting for their own Thought for the Day from Alastair Campbell. Diana Organ, MP for the Forest of Dean, which has been so badly hit, said that she had written to Mrs Beckett on the subject and that she felt it was important that there should be a public inquiry, chaired by a completely independent party who was not connected with government, preferably someone with an independent scientific background.

But these pleas have been to no avail. On 18 July, during the last Prime Minister's Question Time of the parliamentary session and the final appearance of William Hague as leader of the opposition, Tony Blair firmly ruled it out. In answer to a question from Hague, he told the Commons that such a public tribunal was not necessary. There would be a 'different kind of inquiry' and its results 'published'. Hague persisted. The Prime Minister had said there was need for a proper inquiry. Surely, therefore, it must be a public inquiry?

'I don't agree that it has to be a public inquiry,' Blair responded, 'for the reasons given by the Royal College of Veterinary Surgeons, which said it was sensible in order to produce an inquiry quicker, that we should have a different type of inquiry. We will have that inquiry and publish the results of that inquiry.' So far as he is concerned, therefore, that's that.

As for the attitude of the NFU, opinion appears to be divided. The executive is asking only for a 'government inquiry', while for some of the organisation's most senior members that is simply not enough. Ian Johnson, from the NFU in the south-west, pointed out that a 'proper inquiry' is not the same as a 'public inquiry'. 'We need to make sure there is public access to rebuild the mistrust by the farming community.' He acknowledged, however, that this had to be balanced with the need to get some answers reasonably quickly but that there were ways of doing this.

However, the prize for government arrogance, coupled with flippancy, must surely go to Agriculture Minister Elliot Morley when, on a visit to the National Scrapie Centre in Worcester on 20 July, he told the *Worcester Evening News* that the government were not going to hold a public inquiry since it would 'attract every crackpot in the country to air their pet theories', a comment better left unsaid which is still ricocheting around the country and is likely to return to haunt him.

What he did not say was that a public inquiry, which has full legal status, would have the right and power to summon Ministers and that the Minister who 'took control' of the epidemic back in May was the Prime Minister himself. Theoretically, therefore, Tony Blair could be asked to explain just what he did and why he did it. Also, it would have the power to summon the senior civil servant, Bran Bender, who was in charge during the outbreak and who had been rewarded by being put in charge of an even bigger empire.

There is also another point that needs to be made. We hear a lot these days about how the country is developing a 'blame culture', that people must find someone, something or some organisation to blame every time things go wrong. This premise has increasingly been heard to describe public reaction to events as diverse as the Southall, Paddington and Hatfield rail disasters, the disaster of BSE and the Bristol heart unit scandal and it is increasingly being used in connection with the foot-and-mouth epidemic. But those who glibly put it forward are rarely, if ever, asked, if that is indeed the case, why should this be? To which the answer is that if we do indeed have a 'blame culture' it is because alongside it we also have a 'cover-up' culture and an 'I/we weren't responsible culture' too.

There are obviously many questions people would like answered, but these are some of those mostly commonly asked in all the areas which have been affected by the epidemic:

1. Foot-and-mouth is endemic in many parts of the world. So why was there no contingency plan in place to deal with a possible outbreak in the UK?
2. When did the epidemic really start? At Heddon-on-the-Wall or, as some of those involved now believe, in the autumn among sheep flocks in southern Scotland?
3. If it did start earlier, then why wasn't it spotted?
4. If the first outbreak was at the Waughs' farm − and certainly improperly treated pigswill has been a major source of infection in previous outbreaks − why was so little notice taken of those people who complained of conditions at the pig unit? Not to mention why action was not taken when witnesses reported seeing limping and dead animals?
5. Once it was confirmed at Heddon-on-the-Wall, why did it take so long to stop all animal movements?

6. If slaughter was the preferred option from the first, then why did MAFF not shut all the marts immediately, send staff in for animal welfare purposes and then slaughter all livestock in them immediately, before these animals spread the disease all over the country? Then, having shut down all the marts at the first moment of diagnosis, trace the movements between them to see how many animals had been affected, and slaughter them? As this was not done, the slaughter policy, from the beginning, had a great big hole right in its centre. What is the point of chasing after the virus with a slaughter policy when the animals had already been let out?

7. We are told the government and MAFF did not realise the scale and complexity of animal movements at any one time. Why not?

8. Faced with what was obviously a serious outbreak, why were the recommendations of the 1968 Northumberland Report on the 1967 outbreak not taken on board?

9. Apropos of that, why did it take so long to call the army in?

10. It is not just 'crackpots' who believe we should consider a vaccination policy, at least alongside that of slaughter (as happened in Holland which is now clear of the disease), and many international experts could be called to put the case for it. So why has so much misinformation about vaccination been peddled, some of it described by knowledgeable critics like Anthony Gibson as outright lies? Furthermore, why did MAFF give out erroneous information as to what EEC policy actually was? Was it through ignorance or was it deliberate? Were they fully aware of the efficacy of new vaccines? Those involved appear to have been arguing on the basis of 30-year-old science. If they did not know the up-to-date facts, then they should have made it their business to find out. If they did, then they were, at the very least, being economical with the truth.

11. Why did the government and MAFF refuse to listen to any but their own experts, turning down offers of help from those who had experience of the 1967 outbreak and advice from a whole raft of specialist academics – not least the Pirbright scientists?

12. Why did the government appear to take the interests of the farming

lobby more seriously than those of the broader rural economy? At least that's how it appears to those involved in the tourist and holiday industries.

13. Having gone for slaughter as the only policy, why did MAFF not ensure that there was sufficient equipment and personnel in place to complete the procedure within the crucial 24 hours after infection had been confirmed?

14. Worse, having further embarked on the policy of the contiguous cull, on the basis of the Imperial College computer model, why, in the light of recent experience, did they again go ahead without ensuring there were sufficient resources to carry it out effectively, along with a massive programme designed to ensure fast disposal of the resultant carcasses?

15. How reliable were the subsequent tests on livestock? There is a great deal of evidence to suggest that they were not reliable at all. Tests carried out on livestock finally culled, after delays long enough to allow them to show symptoms of the disease had they been infected, were later proved to have been negative.

16. Is the virus carried on the wind? There are totally opposing views on this.

17. What precautions were taken to avoid subsequent public health problems that might be caused by smoke from the pyres? What monitoring was carried out with regard to the constituents in the smoke and how soon was this done?

18. Why were EEC pollution regulations ignored with regard to the pyres?

19. Was it possible the virus could have actually been spread by the pyres?

20. Has the law been broken as regards the contiguous cull? Judgements have been confused. There are certainly grounds to believe that this happened in some cases.

21. What contingency plans are now in place to cope with another outbreak?

To these might be added other questions which have arisen at the end of July. Why has it taken so long for the government to discover the costs involved in disinfecting farms? Why does it appear that contractors were offered open-ended contracts? Why was the cost of such work being carried out by farmers themselves, with some help, not compared with that of calling in the big boys? And why, if biosecurity is so important, can such disinfecting be put on hold while the financial side is sorted out?

David Hill, chairman of Devon NFU, who farms near Beaworthy in west Devon, puts the case for a public inquiry from his own perspective and has his own questions which he would like to be answered. Unlike some of his colleagues further north, he feels that the initial response to the outbreak was quite good but within days the pressure showed and the response rate diminished in proportion to the increased thrust of the disease. Not only was the army brought in too late, in Devon at least, it was impossible to sort out who was running the show.

One of his first questions concerns the organisation, if it can so be called, in the south-west region and who was, supposedly, in charge. 'From where I sat, both the army and MAFF tried to claim the high ground, with the inevitable chaos. There are plenty of farmers who know that MAFF would order vehicles and the army cancel them or vice-versa. Similarly, pyres were built and then the lorry would arrive to take the sheep for rendering.' Sleepers would arrive but not straw, lorries were sent to enter lanes they had no chance of getting down. Slaughter teams would be on site with no vet. Vets without valuers. And how do you slaughter 100 animals with 99 bullets? All that happened. Why?'

Other questions he feels need putting to a public inquiry would include: How was compensation arrived at? Why was it changed part way? How can you justify the same money for a Welsh ewe as a good Suffolk? Or for old cows against prime breeding stock? Then there is the question of the great Ash Moor pit. The sum of 'a dream lottery win to dig three holes using taxpayers' money all the way, like the rest. Was it needed?' Was £7.5 million the correct spend? And landfill sites – are they safe or not?

He then returns again to the question of the contiguous cull. 'It surely remains the folly of this whole sorry saga.' A Minister had kept claiming that the outbreak was under control when certainly action was needed. But what kind of action? The election had made immediate action essential and 'what luck that there should be a body of scientists on hand to recommend mass slaughter. Kill 'em before they get it, was the theme. So they did. With military effectiveness. Hundreds of thousands of animals, some over a mile from the infected place . . . until, suddenly they didn't. That is worth an inquiry in its own right. Were there second thoughts? Was it the 150,000 carcasses putrefying in the Devon countryside, or was it that the nightly news items were increasingly horrendous? Or did the deadly hand of the Treasury descend? We should be told.' (At the time of writing, while there have been no new outbreaks in Devon for four weeks, the contiguous cull – unreported and unremarked upon outside those areas infected – goes on in Wales, Yorkshire and Cumbria.)

Hill admits that it is almost too difficult to know where to start looking for the reasons for the whole sorry affair, 'but open government, transparency and accountability are some useful pointers, and those who have been devastated by the disease would put "truth" as an overwhelming need, with "responsibility" immediately behind . . . And perhaps therein lies the key to prevarication which has typified the new Ministers' utterings on this matter. For responsibility, good old-fashioned "who was to blame" responsibility, is exactly what we need to apportion. Light into dark corners should be the order of the day and those exposed by it should carry the can – whoever they are, without exception.'

He concludes by echoing what the rural community is saying from end to end of the country. 'And so to the best reason of all for an inquiry. We must never, ever, allow this disaster to happen again. Viewed from abroad the country appeared to be a battlefield, covered in blood and smoke, while killing gangs went about their terrible work. The tourist industry imploded and red meat sales plummeted. The true cost will never be known.'

But who allowed the contingency plan to wither and die? We need one in place, revisited every year and incorporating all the lessons of the outbreak. 'How can that be achieved without knowing what went wrong and what should be put in its place? That information has to be extracted, learned from and recorded. It cannot be unless there is a full and open public inquiry.

'My personal wish would be for county tribunals leading up to a national inquiry. Let those who ran the chaos face those who suffered from it. This horror has scarred our lives for ever. At the very least those people should have the comfort of knowing they did not lose their lives' work to further political ambition. Or did they?'

# Chapter 16

# But It Isn't Over . . .

But while now all the talk is of what should happen in the 'aftermath' of the disease, how the farming community needs putting together again, what can be done to help the rural economy in general and 'learn lessons', the epidemic continues. Not, admittedly, at the rate it did at the very height of the crisis when it was never out of the media, but well beyond what can be described as 'tailing off' or 'the aftermath', whatever politicians would have us believe. Even worse, if nothing is done, if no new strategy is put in place, it is almost certain that by the autumn we shall find ourselves back at square one. It is as bleak as that.

It needs to be remembered that shortly before the general election, Chief Scientist David King reassured everybody that the epidemic would be virtually over by 7 June, which was just what the government wanted to hear. However, with the election out of the way, he qualified that to it probably not being entirely over until August. One doubts that he would risk saying that again or even hazarding a guess as to the date of the endgame.

The total number of outbreaks now stands at (2,007). In the nine weeks up to and including 23 July, there were 44 outbreaks, 26 in new areas. There were 14 new infections during the week beginning 16 July, including the first ever in the Greater Manchester area, and another just outside Wigan. On 9 July 20,000 sheep were killed in Yorkshire as part of the contiguous cull in the area.

On 22 July Christopher Laurence, head of the RSPCA's foot-and-mouth strategy group, told *The Observer* that rates of infection could return to epidemic levels this autumn as the disease once again spreads throughout the country thanks to the government's 'catastrophic management'. He stated that DEFRA's failure to carry out sufficient blood-testing of the nine million ewes roaming free throughout Britain would make it inevitable that sheep carrying the virus would mix, undetected, with uninfected livestock when the flocks were brought down from the hills in the autumn.

'The lack of routine checks,' he said, 'and confused methodology, poor routine and controversial interpretation of test results, gives us little confidence that there is anything stopping the disease spreading like wildfire come the autumn.' Which would start the mass culling and consequent restrictions on farmers all over again. He admitted that DEFRA was now trying to make up for lost time but that it was almost certainly too late. 'They have failed to learn any lessons from the past year and there seems to be no impetus at all to prevent this disaster recurring.'

Richard Sibley, President of the British Cattle Veterinary Association, said he had been lobbying the government since March to step up the rate of blood tests. 'This is something we are all desperately worried about. We need to test 1.3 million sheep in the next six weeks, but even if tests go ahead at full capacity from now on, it will take 14 weeks to get all the tests completed.' Pathologist Joe Brownlie had offered to set up a laboratory to carry out the necessary tests back in April but was told that it was all in hand. 'We could have had the labs up and running within two weeks,' said Sibley, 'it's hard to justify how it's taken another four months to establish what we offered to create all that time ago.'

DEFRA claimed that from the testing it had carried out so far, only 0.16 per cent of the sheep tested had proved positive. The response of Minister Elliott Morley (he of 'crackpot' fame) was that 'these sorts of warnings are totally irresponsible, on the evidence we have so far, we're quite confident that there's quite a low level of unidentified infection, although I'm not saying there's no infection at all. There's a risk when the sheep come down from the uplands, but we're doing our best to quantify that risk.' To which David McDowell, an RSPCA foot-and-mouth adviser to the government, said that it was thought that the true figure of sheep carriers was closer to 10 per cent. 'DEFRA's tests are hit and miss, and secretive to boot; they've refused to make public how they're going to interpret the results and we're not even certain what tests they're doing.'

At the end of July, 4,000 sheep were killed on the Brecon Beacons in spite of enormous opposition from the farmers involved. Only 112 of them were shown to have had antibodies in their blood. This time the item did make headlines, at least it did on the *Today* programme on 27 July when a shepherd pointed out that the sheep had tested positive for antibodies not the virus, which could mean they had had the disease and survived it or been in contact with it but not succumbed to it. Virologists are in general agreement that, if this is the case, then the sheep with antibodies pose no threat. To say that they do flies in the face of all the latest research on the subject.

Among others interviewed was Julia Phillips, who was about to go out and round up her sheep for the slaughter. She said that she felt she had to do this herself, since DEFRA slaughter teams had no skills in dealing with frightened animals. She too emphasised that the sheep about to be killed were healthy sheep, only a handful of which had even shown positive for antibodies and these could have been killed, if it had been thought necessary, and the rest which were clean left to live. Like all the other farmers involved, she said, she was given no option but to give in to the slaughter. Vaccination was never even discussed and DEFRA had, as had the old MAFF, frightened people into agreeing because they could not afford to risk the legal costs if they went to court and lost.

'There used to be a thriving Graziers' Association here,' she said, 'but it's been broken up. We've been picked off one by one. This will be a horrible, devastating day. I'll now be going out with my dog to round up my sheep knowing it might well be the last time I'll ever do it.'

The programme also interviewed Ken Tyrell, mentioned before as one of those involved in the 1967 outbreak. As well as making scathing criticism of the way the government had handled the crisis from first to last, he also expressed concern that the way the government and DEFRA have chosen to dispose of carcasses may well be spreading the virus, not containing it. He pointed the finger at rendering plants as a possible source of infection in themselves, not to mention the wagons travelling across the country full of infected carcasses. As for the contiguous cull, it was almost as if the old MAFF had been swept by some awful kind of blood lust. 'MAFF wanted a clean sweep: empty fields don't have the disease.'

In response the government fielded Chief Vet Jim Scudamore, now a rare visitor to the airwaves. He spoke of the government's achievement in ridding most of the country of the disease 'except for three or four hot spots'. When it was pointed out that immediately before the election Tony Blair had said that we were on the home straight, and the Chief Scientist that by 7 June there would be 'nil' cases a day, he responded that he had always said there would be a 'long tail' in the aftermath of the epidemic, and that this is what we are now in.

John Humphrys again returned to the question of the mass contiguous cull, of the hundreds of thousands of healthy animals that had been killed as a result and commented that the policy, from the start, had surely been driven by political considerations, not farming or economic concerns. To which Scudamore replied, robustly, that this was simply 'not true!'

The following day, BBC radio returned to the subject again. The cull had been delayed for hours, due to the opposition of farmers – put down to arguments over valuations – but was going ahead. Another farmer interviewed again brought up the question of it being antibodies that had been found in the blood, not the virus, and that there was simply no reason to cull thousands of healthy sheep. They were all in pens, out of the way of other livestock and could be left there for two weeks to see if they would develop the infection. He made the point again that these were irreplaceable 'hefted' sheep, and that special measures were supposed to be applied.

He also drew attention to the shambolic nature of what was going on. Healthy animals were being culled willy-nilly, and the relevant farm premises scoured to the stone with disinfectant, whereas possibly infected sheep had been slaughtered and not only had disinfection procedures been minimal, but the pens in which the culls had taken place had been left for days with pools of blood on the floor along with bits of flesh and bone. This was not only disgusting and showed just what a cull entailed, but was an invitation to scavengers like foxes and crows to pick at the detritus and, if it was infected, carry it elsewhere.

An appeal to Elliott Morley on the same programme fell on deaf ears. This was the way to deal with it and this was the best advice. Pressed on why the Royal Veterinary College had received no reply from Agriculture Minister, Margaret Beckett, to its letter (sent before the parliamentary recess) asking for a full and open inquiry as soon as possible, Mr Morley replied that this decision was 'for the Prime Minister only'. But surely we need to have a public inquiry to look into every aspect of what had happened? No, Tony Blair will decide what kind of inquiry there will be, when the epidemic is over. Yes, there will be an enquiry. 'We need to look at what has gone right, as well as not quite so right . . .' A statement which left the interviewer virtually speechless.

The thousands of sheep slaughtered may well herald an even vaster cull, talk is now of a further 6,000. These are hefted flocks, such as those in Dumfries and Galloway, the Borders, Cumbria and other areas and, as such, were supposed to be protected from unnecessary slaughter but the promises made before the election now seem hardly worth the paper they are written on. In the Royal Forest of Dean the hefted flocks have been completely culled out. The 'divide and rule' ploy has also been used elsewhere. In the Golden Valley in Herefordshire, villages are so small and isolated that they can easily be closed off. The sheep have also gone from Sugar Loaf Mountain and are now likely to disappear from the Black Mountains.

Meanwhile, farmers in Cumbria are sour that all the attention has now turned on Wales, when between 1,000 and 2,000 animals a week are still being slaughtered in their county without anyone apparently taking any interest in what is going on.

Farmers' leaders say they are facing a disaster which is rapidly turning into a tragedy. The disappearance of the hefted flocks will have a disastrous effect on the landscape and wide-ranging social and economic effects on the whole community. This year's holiday trade has already gone, next year's looks like following suit unless there is a complete change of policy and that means that vaccination must be considered.

But the slaughter policy was, Scudamore insisted, working. So the slaughter goes on. DEFRA now plans to 'take out' the whole of the Aire valley from Skipton to Bradford/Leeds and has slaughter teams and vehicles in place to do it. Farmers there have been told that they are to blame, the infection having been carried on vehicle wheels, although there is no proof that tyres carry the virus.

But Scudamore could be whistling in the dark, with his talk of a 'long tail' and 'three or four hot spots'. Already big pits are being dug at two disused aerodromes in East Yorkshire because of the very real fears that the virus will spread to the massive pig units in the area and from there into Lincolnshire by the end of August.

The buzz word for the government from the beginning of July onwards has been 'biosecurity'. Scrupulous attention to detail by farmers would be needed to ensure the epidemic was finally brought to an end – everything must be done to control any possible spread of infection – and it was to this end that a video on the subject (already referred to) was sent out to farmers. On every possible occasion the message was brought home that the government had done its part and that now it was up to farmers to ensure the disease was contained.

But the 1968 Northumberland report had already pointed out that this was very difficult to ensure because of the very nature of farming, however conscientious a farmer might be. Hill farmer Richard Mawdsley agrees. 'If and when all tested premises test negative, then D notices will typically be lifted quickly and the Infected Area boundary rolled back. Sadly the virus makes no concessions for people being 'only human'. If effective biosecurity is not maintained and there are movements on and off the farm, then the virus can and will continue to spread.

'I am increasingly of the opinion that it may be unrealistic – at least it's beginning to appear that way – to expect people to establish and maintain effective biosecurity for long periods over a large area, especially when in some cases people/farms who have not had the disease are now in worse financial circumstances than those who did. If that's the case, then the policy of containment by means of movement restrictions and biosecurity, followed by eradication by slaughter, cannot work effectively. In which case vaccination becomes the only effective solution: that or just "giving up" and letting the disease run rife.'

An increasing number of cases, he points out, are now being picked up on patrol visits rather than via farmers' reports. This suggests inadequate inspection of stock, 'or, heaven forbid, deliberate failure to report suspicions'. Inadequate inspection of stock and failure to report, combined with ineffective/inadequate biosecurity is a recipe for disaster when using the current disease control policy. There have recently been rather a lot of cases outside existing protection zones – so called 'sparks' that then sometimes go on to form new clusters, which has prompted protection zones to be expanded from 3 km to 10 km, and potentially further should the need arise.

He speaks of the growing fears that the virus will spread into the dense pig populations of Humberside and East Yorkshire, 'towards which it is creeping, hopping, ever nearer, where there are 730,000 pigs, 128,000 of which are held on premises that also have cattle and sheep. Whilst pigs are harder to infect than cattle or sheep initially, they do emit many, many times more virus than cattle or sheep, and the way they are managed exacerbates the viral load generated, so infection spreads very rapidly in dense pig populations, and risks spiralling out of control very quickly.'

It is against these fears, this background and the publicity given to the need for the utmost biosecurity, that the government exerted maximum pressure to have most rural footpaths opened to the public again while, at the same time calling a halt to the disinfecting of farms on the grounds of excessive expense, costs rising from £30,000 per farm in Scotland to £100,000 in England. At first it was said that stopping the disinfecting while costs were looked into would apply only to those farms yet to embark on the process, within hours there had been yet another turnaround. It applied to all farm disinfecting at whatever stage the process had reached, even if it was halfway through.

No one is arguing about the fact that some contractors have made a killing (because no controls were put in place), or that DEFRA's orders have varied between one area and the next, or even one farm and the next, or that some of the lengths farmers have been required to go to have been ludicrous. To return to Ken Tyrrell again, he says he was dumbfounded at MAFF/DEFRA ignorance of how the virus behaves and the degree to which the clean-up has got out of hand.

'The virus has not changed so why have the cleansing and disinfection methods been allowed to reach these completely over-the-top limits? In 1967–8 we had 2,346 outbreaks and in only 18 cases was there a case of reinfection due to infected material remaining on a farm. Cleansing should be much faster, which would save money and help get battered rural lives back to some degree of normality.'

Confidence has not been enhanced by Elliott Morley's statement, in yet another BBC interview on 26 July, that DEFRA 'is in control of the process', with all the connotations carried by that particular phrase. If standards are so diverse and so varied, then it patently is not. But the effect on the ground has, once again, been a complete shambles. David and Nicky Hayes of Clayhanger in Devon (one of the hot spot areas now clear of the virus), were halfway through cleansing and disinfecting their farm when they were ordered to stop all work until further notice. They had been told they would be paid by DEFRA to carry out the job themselves – they have so far been paid nothing. They are now in limbo, having hired expensive equipment which they cannot afford unless they are allowed to use it and claim for it. They lost over 600 lambs and ewes and 30 cattle in a contiguous cull. 'We were just beginning to think things might be OK and then this happens. We just don't know what to do for the best anymore,' said David Hayes. 'It will take at least another six weeks to complete the process once they have been given the go-ahead to continue.'

Another farmer, Anne Heard, who lost 1,000 sheep and 300 dairy and beef cattle was in the process of being disinfected by a contractor when she had a phone call from DEFRA informing them that all work would be suspended while a review of the cost was undertaken. That was, she said, after she had heard on the radio that since their disinfecting process was already taking place, they had nothing to worry about. She was, she said, 'absolutely stunned', when she got the phone call. Things were going well and they had hoped to get something ready for the harvest. Work stopped at once and many of the farm buildings lie in pieces where workers have been dismissed and have walked away. They have no idea when it is all going to be put back together again. Many of those helping the Heards were farmers and labourers who had nothing to do as they had been 'slaughtered out', and DEFRA was running around six weeks behind schedule in paying them.

Anthony Gibson agrees that the cost of disinfecting farms had got out of hand, giving some extreme examples of farmworkers having to crawl 'like rats in a sewer' beneath the slats in a cubicle house to pressure-wash the sides and ceilings of slurry channels already disinfected with citric acid, 'and this on a farm which never actually had the disease'. He shares other stories from Wales of contractors chasing vets like American lawyers chasing ambulances and of contractors turning up on farms before the dead livestock is cold, offering their services on the grounds that they have a good relationship with the supervising vet. The answer he says is first to ensure greater consistency from one farm to the next by using standard guidelines based on a risk assessment; and the second is to supervise the contractors properly – not stop the process in its tracks.

There is no doubt that the sudden and unexpected about-turn on disinfecting is sending very confused messages to farmers. Another farmer, Andrew Gunther, spelled out the complexity of the disinfecting process, every land drain, barn, trough, all fittings, had to be cleaned and sprayed, yet the rules were changed every day. He was told floors had to be taken up, then that they had not, big holes in walls had to be filled in but not small ones, walls had to be washed, no, they could be lime-washed (cheaper and quicker). The disinfection officer had 70 farms to supervise, telephone help-line numbers were changed, and money was continually being paid out in large amounts but farmers' invoices remained unpaid. It took a great deal of heavy pressure to extract at least some of the money out of DEFRA.

'In all,' he says, 'I have made 60 calls to chase invoices and I expect to make a load more. Please, Mr Blair, see for yourself. You took personal charge of the FMD crisis – surely you could make a better job of this than the mess it currently is?' He says he does not blame his neighbours (which is why he was contiguously culled), or even the MAFF/DEFRA officials on the ground. 'It's the complete incompetence of the system and the lack of decision-makers.

'In the end I blame Mr Blair. I wrote to him and received a reply from some secretary. Despite his reassurances it was just rhetoric. If he had talked to us or talked to his staff, we could have saved the public's time and money and we all know what this is costing. I suppose he forgot to keep his eye on the countryside while trying to save the taxpayers money. The Dome cost £600 million and had little or no effect on tourism. If foot-and-mouth gets going again it will cost billions and have every effect on tourism.'

The crunch time for knowing whether or not foot-and-mouth is going to romp away again will be around October when the weather gets colder and, as is likely, wetter: the conditions in which the virus thrives. That the government itself might not believe its own brave words was revealed in a report by Christopher Booker in the *Sunday Telegraph* on 15 July, in which he says that he has learned from a well-placed source, that DEFRA now has 1,600 vets under contract (700 more than MPs were recently told was the case by Margaret Beckett) and most contracts are due to run until May 2002. The vets receive £260 a day, plus £22 meals allowance, even though they have free board and lodging, including breakfast and dinner, and a further £20 a day 'stress allowance'. This equates to £78,000 a year, which, as Booker points out, was until five years ago the salary of the Prime Minister.

In Holland now they have no such worries, the epidemic is behind them. The day after the discovery of the virus in the Essex abattoir, Holland banned all imports from the UK and all livestock markets were stopped. Animals were culled on all farms with imported UK livestock and there was a preventive cull of two sheep farms. On 12 March, when a case was identified in France, Holland banned all animal transports. On 19 March, the Dutch Agriculture Minister asked the EEC for permission to vaccinate. Two days later cases of foot-and-mouth were confirmed in four cattle that had been transported from Ireland via France during which time they had come into contact within infected stock from the UK. Within four days there were 25 infected premises and the animals were first vaccinated, then slaughtered. All livestock owners within infected areas were told to have their animals vaccinated.

On 23 April the EU allowed the resumption of exports of fresh meat, meat products, milk, milk products – but not live animals – from selected areas. On 8 May, meat exports from Holland resumed excluding those from infected areas. By 22 May, screening of animals had been completed and all were pronounced clear. Six days later, all restrictions on the recreational use of the countryside were lifted and by June those on transportation for slaughter and all-over restrictions were lifted.

The Dutch policy is that as soon as a case of foot-and-mouth has been confirmed, all susceptible animals within a two-kilometre radius of that farm must be tested, after which emergency vaccination is used so that the virus is unable to spread even if the culling of all animals inside that two-kilometre ring, and their disposal, cannot be carried out immediately. Altogether the Dutch culled 200,000 animals and estimate the cost to the economy at 2.8 million Dutch guilders (approximately £800,000). The Dutch are now pressing for preventative vaccination but this would have to be available to everyone, otherwise it would put the country at a trading disadvantage. But the advantage of using emergency vaccination, even if animals were culled later, should be plain enough for anyone to see. Holland's exports will be up and running months before those of the UK.

Halfway through June, Ben Gill told the press that the NFU had 'made some mistakes', but that the refusal to allow vaccination was not one of them. In spite of all that had been written, said and proposed; despite the fears of a flare-up in the autumn, he insisted that that 'at the end of the day with foot-and-mouth, there is only one policy'.

Professor Vernon Wheelock, former head of foot-and-mouth policy research at Bradford University, and now running his own consultancy near Skipton, told the *Yorkshire Post* on 18 July that the opposition of the NFU to vaccination was 'a misjudgement of monstrous proportions'.

Unable to get any response he then sent an open letter to Ben Gill detailing the present position as he sees it and stating that, contrary to what we were told and what is still being said, the number of new cases is not falling. 'Clearly the policy isn't working.' He also expresses great concern at the intensive activity in selected hot spots, such as Skipton, Settle and Clitheroe, where 100 farms have been culled out and where DEFRA and army teams have been extremely active. He thinks there are compelling reasons for believing that the teams themselves have been responsible for helping to spread the disease and that he had evidence of failure to adhere to strict biosecurity rules.

He points out that with the warmer weather it is correct to expect that the virus would have difficulty surviving, but for whatever reason, that does not appear to be the case – it is still rampant. 'Unless it is brought under control within the next couple of months, we will be going into the winter with the strong possibility that it will still be present this time next year.'

He blamed the government for giving in to NFU pressure, pressure that he finds hard to understand. He thinks it 'beyond incredibility' that the big supermarket buyers and major retailers would not buy meat from vaccinated animals, as they already sell such meat from sources outside the UK. Therefore if some arrogant buyer or retailer refused to buy British on the grounds that it came from vaccinated animals they would be laughed out of court.

He lists four of the experts who press for vaccination, Dr Fred Brown FRS, Dr Paul Sutmoller, Dr Simon Barteling and Dr Paul Kitching, as well as all the work which has been undertaken by virologist Dr Ruth Watkins at St Mary's Hospital, London. For Mr Gill's benefit he lists again the facts about the foot-and-mouth disease vaccine: that it is a killed (i.e. inactivated) virus; it induces immunity without infecting the vaccinated animal; it is a type used for 40 years and is well tested; virtually 100 per cent of animals respond to it; it is licensed for use in the UK; protection is confirmed after one dose; the vaccine provably protects against the disease and the vaccine will not infect others, whether they have been vaccinated or not. The transmission of the virus is halted and vaccinated animals can prevent infection after restocking. All meat products, milk and milk products are safe to eat and drink.

'If a decision to vaccinate was taken today, foot-and-mouth disease would be history by September.' If he [Mr Gill] would change his mind he would have taken a step, Wheelock tells him, 'towards regaining some of the respect you have lost, but if you maintain your current position you leave yourself wide open to be castigated when the inevitable inquiries eventually get down to work'.

'Furthermore,' warns Wheelock, 'when it is eventually recognised that the use of vaccines could have avoided most of the recent culling and all the trauma that goes with it, the government will be only too keen to look for a scapegoat. Do we need to say any more?' It was vital that the NFU should extract itself from its bunker mentality and help get the government off the hook. He recognises that it would be a big hurdle to overcome, that 'many people, including yourself, would have to admit to having made a fatal error of judgement in ruling out vaccination months ago. This is the challenge. Are you big enough to accept it?'

Cumbrian vet Jim Clapp says that what will happen next without vaccination is that foot-and-mouth will become endemic in the UK. He considers that to be enough to make it necessary – endemic means that it will spring up again at any time and in any place forever, and, of course, exports will never be allowed again. Dr Ruth Watkins states that there is only one way of getting rid of an asymptomatic contagious disease and that is vaccination. (Asymptomatic, because it is passed on before you know the animal has got it.) Other critics point out that it has now reached the stage where there is no longer even any point in killing all the livestock in the UK, as it is endemic in wildlife.

Writing in his regular farming column in the *Western Morning News* on 25 July, Anthony Gibson said he had bumped into Lord Henry Plumb a few days earlier, and that the great man's legion of admirers would be happy to know that he is hale, hearty and as interesting as ever on the subject of foot-and-mouth. Lord Plumb was chairman of the NFU's Animal Health Committee during the 1967 outbreak and is the only surviving member of the Northumberland Committee.

'I asked him which of the many lessons of that inquiry had been most culpably ignored. "Oh, that's easy," he replied. "We warned them: either you do something to control your meat imports or you will have to move to a vaccination policy. Simple as that. And it still is."'

As we move towards the autumn with new cases still being

confirmed almost every day, Ben Gill still refuses to change his mind. It will be interesting to see, when the costs of the epidemic are finally brought home to Gordon Brown, if the Treasury will not then step in and do it for him.

# Epilogue

Writing this book has been a race, not only against time but against fast-moving events, contradictory government statements and activities, and a situation which changes ... by the month, week, or even by the day. The foot-and-mouth epidemic, the death knell for which was prematurely sounded in early June, continues even if, as government spokesmen tell us, it is now only 'confined to hot spots'. It is August and mass slaughter is still the sole policy as the disease continues to spread. There is a strange sense of *déjà-vu*. The story began with inefficiency, ineptitude, lack of proper planning and panic-stricken attempts to stop a disease which has always been one jump ahead. Unhappily little has changed and not much learned. Thousands more sheep are likely to be slaughtered as it spreads ever further into the Welsh hills, an attempt at control which might well be in vain if pessimistic forecasts are true and it is now endemic in the hill flocks.

North Yorkshire sounds like a war zone. After all the messages on biosecurity and the white-suited slaughter teams rushed into Skipton and Thirsk, the government is finally coming to grips with what its critics have been warning of for weeks: that these very slaughter teams, contractors and wagons may well have been helping to spread the disease as they go; that biosecurity has been sloppy; bits of infected carcasses have been left lying around for the crows to pick and foxes to eat, and that contaminated wagons have been travelling the roads. One might reasonably ask, therefore, what has changed since February?

Terrified of the virus reaching the massive eastern pigs units, police and trading standard officers are now patrolling an area 600-kilometres square in a desperate attempt to prevent the worst from happening. It must be holiday time for lawbreakers. DEFRA officials are described as riding 'shotgun' on licensed milk and feed lorries to ensure the vehicles are properly disinfected, following the decision to slaughter 9,000 pigs at Skipton-on-Swale, near Thirsk, because the breeding unit there might be a 'dangerous contact'. 'Alarm bells are ringing,' we are told. Really? After six months of a continuing epidemic?

Foot-and-mouth is also back in the headlines after nearly three months of almost complete silence, during which most of the population assumed it had gone away. Perhaps the words 'conspiracy of silence' are too strong, but every effort appears to have been made to keep the bad news out of the papers and put a positive spin on the rest. It will have become apparent through reading this book that there is a lack of response from MAFF/DEFRA to specific questions. This is because it has been virtually impossible to get straight answers to anything at a national level. The enquirer is merely referred to the relevant website from which, during the weeks of the election, even basic information disappeared.

However, in spite of all the lack of official news, there has been an abundance of material available for those interested in seeking it out. I began this book by quoting Jon Snow when he asked if the foot-and-mouth epidemic was becoming the first e-mail democratising story of our time. The answer to that is 'yes'. It has been possible to discover news of what is happening on the ground very fast; ask for help and receive it within the course of a day and exchange information and compare notes.

The many relevant websites also have provided an enormous amount of information. It has been possible, for instance, to download the 1968 Northumberland Report or a précis of it, along with literally dozens of academic papers, many on aspects of vaccination, others on the history of such epidemics in general and foot-and-mouth in particular; to discover what has happened in Holland and how, by the use of vaccination, it is now clear of foot-and-mouth disease; to learn that the retailers who say they would not stock meat from vaccinated animals because of consumer resistance, regularly buy it in from abroad and say nothing. It has been very difficult for the authorities or individuals to cover up bad practices of all kinds as, in almost every instance, someone somewhere has witnessed it, logged it and sometimes recorded it on camera as well.

Finally, belatedly, the horrendous economic costs are starting to come home, resulting first in the temporary halting of the disinfection of farms, though presumably it finally struck someone that this might turn out to be a false economy. The scheme was restarted again, amid major disruption, after a fortnight.

Next, the government announced it was moving to curb the 'abuse of the compensation scheme'. So far over £1 billion has been paid out on this alone. In future there will be no more valuing by private valuers, possibly in cahoots with a farmer, but it will be carried out by valuers chosen by DEFRA. This sterling declaration was, however, somewhat undermined by the information coming out within days that 37 farmers had become millionaires on compensation payouts: one, in Dumfries and Galloway, received £4 million. While the ordinary citizen was reeling from this news, it was revealed by the media on 26 August that those farmers who had insured against foot-and-mouth, had been paid twice for culled stock: once by the government in compensation and again by their own insurers. Some of the amounts paid out run to hundreds of thousands of pounds.

There is absolutely no doubt that for the small or medium-sized farms, the epidemic has been an unmitigated disaster. Certainly in Devon there are farmers who have been reduced to claiming benefit and applying for money from the various charities set up to help them. But it's been very different for the big boys represented by those who are described as 'leading figures in the NFU'. 'This double compensation,' wrote environment correspondent Anthony Browne in *The Observer* on 26 June, 'comes on top of excessive amounts paid to farmers because they colluded with valuers to deliberately over-estimate the value of their animals.' The NFU justifies the double payouts by saying that the money from the government covers the market value of the herd, while the insurance money covers loss of income. But Professor Peter Midmore told *The Observer* on the same date: 'The productive potential in livestock is included in the price. Farmers could be getting more compensation that they need.' This news released a flurry of leaks to the effect that both the national audit office and the Commons' public accounts committee, were becoming seriously anxious about both the persistence of the outbreak and the scale of compensations payments.

Thousands of sheep in the Brecon Beacons have now gone for slaughter, amid more allegations of deliberate infection; real criticism over what has been going on has finally begun to surface again in the national media – the regional media has been on the ball all the time. David Rose, again writing in *The Observer* on 5 August, under the heading 'A Dangerous Liaison', drew attention to the virtual terminal devastation of the Brecon Beacons and noted that we, the public, have 'to pinch ourselves to believe that incompetence such as we have seen does not belong to some dark dream . . . A minority sectional interest has not been allowed to usurp government policy to such an extent since the 1970s, when the bosses of the TUC swilled their beer and sandwiches in Downing Street.' Rightly, he had misgivings about the nature of any inquiry the government might hold.

> The greatest danger is that the seamless web which stretches from the NFU, through government's scientific advisers, to the ministeriat, will ensure a fix: an inquiry predetermined to conclude there was no alternative to what was done, however terrible its consequences. That the evidence heard by the handpicked inquiry team will never be tested by public cross-examination, and if it should be given cogent reasons why the whole policy was a washout from the start, then that can be buried in the Public Record Office for another 30 years.

He proved to be right on target. On 9 August Margaret Beckett announced the government's decision. There are going to be three inquiries, as well as the host of smaller ones covering other aspects which will bring the number of 'inquiries' into the epidemic to ten, none of them public or legally binding on witnesses to appear. Tony Blair, said Mrs Beckett, would no doubt want to make his own views known, but is 'unlikely to give evidence in person'. She made her announcement while on holiday in France but from the next statement it appears she must have spent the last 12 months on another planet – the officials' control of the disease has, apparently, been 'a howling success'.

The three official government inquiries will cover the future of farming, the handling of the crisis and the science of livestock epidemics. The most controversial appointment is that of Sir Donald Curry to chair the first. It was he who chaired the meat and livestock commission during the BSE crisis, later heavily criticised by the consequent public inquiry for failing to meet its statutory objectives and for 'hyperbole displaced accuracy' (quite a phrase) in some of its campaigns on meat safety.

The chair of the second, Iain Anderson, has worked closely with Mrs Beckett before, and was her special adviser on the millennium bug, for which she had responsibility as Leader of the House. He was appointed to the Board of British Telecom in 1995 and is currently chairman of the board of BT Scotland.

Third and last, Sir Brian Follett, a distinguished zoologist and Vice-Chancellor of Warwick University, will be chairing the inquiry looking into the science of livestock epidemics, which should involve comparing the policy of mass slaughter with that of vaccination.

It is clear that none of these will meet the criteria of a public inquiry. North Cornwall Liberal Democrat MP Paul Tyler, the Shadow Leader of the Commons, said at once that what was being proposed was 'outrageous'. As he told local television news, what we were being offered was 'an incestuous investigation that will satisfy nobody, and merely make both farmers and taxpayers more suspicious'. He, among others, demanded to know how open the inquiries would be, especially that chaired by Iain Anderson. With breathtaking arrogance Mrs

Beckett responded that she would be 'surprised' if Dr Anderson 'chose' to publish the evidence given to his inquiry 'since this was not normal practice'. There is also considerable anxiety as to who will be called as witnesses to any of them. As matters stand it will be quite possible to ignore anyone criticising anything in any way.

It is hardly surprising that this has been greeted with a storm of condemnation. Regional papers in the West Country, Wales and Northumberland have banded together, running petitions for a full-scale public inquiry, but they are likely to be on a hiding to nothing. The government's mind is made up. Ben Gill, speaking for the NFU, at first seemed to have his own doubts, telling *The Guardian* on 10 August that 'we have ended up with a semi-judicial inquiry that asks lots of irrelevant questions, costs a fortune, takes an eternity to report and tells us nothing'. Yet five days later, at a packed meeting of 350 farmers in Cumbria, he appeared to have changed his mind, agreeing with the government that the proposed inquiries would get the answers more cheaply than a public inquiry and that he had been assured that parts of each inquiry would be held in public.

The story is by no means over. The number of infected premises has passed the 2,000 mark and new cases are still coming in at around three a day. Northumberland is now back to square one or worse, the disease having started up again near Hexham in an area that had been free of it for three months or more. Within days the number of cases had risen to 17 and it is still growing, nor are all the new cases close to each other.

Drastic measures have been brought in to 'contain' the epidemic (again), with the area divided into 'blue boxes', in which the most stringent disinfection precautions have to be taken: feed wagons delivering to farms and milk tankers have to be accompanied by DEFRA officials and licences for fodder and slurry must be obtained in order to move fodder or slurry off farms. Joint DEFRA/police patrols monitor the roads and the army has been called back to help deal with the crisis. On 2 September, in the teeth of strong local opposition, the burial pit at Tow Law was reopened to cope with the increasing number of carcasses as farms contiguous to those infected were culled out. This, in spite of assurances from DEFRA some days earlier that it would not be necessary as the carcasses could be dealt with by rendering.

Once again there are pleas for vaccination, one suggestion being to introduce small pilot schemes in selected areas. Cumbrian vet Jim Clapp has wanted this for months and in a letter to Chief Vet Jim Scudamore, he pleads to be allowed to run such a scheme before the damage to the county becomes irreparable and famers leave, never to return. No special teams are needed, he writes, the farmers could do it themselves to avoid possible contamination from outside teams. He could organise it himself. Simple record keeping, using the legally required ear tags, would keep track of vaccinated stock. Animal welfare problems would be solved and people would be free to go about as before.

On 2 September, the *Independent on Sunday* revealed that the government, under increasing pressure, is drawing up secret plans to start vaccinating before sheep are brought down from the hills for winter. Five million animals are likely to be moved during the autumn. Apparently Gareth Davies, a member of the government's own scientific committee, and former head of the epidemiology unit at the official Central Veterinary Laboratory, was asked to start drawing up plans in July. The Committee has set itself until the end of September to finalise them. The paper claimed that the government was prepared to authorise limited vaccination within six weeks if the disease has not been eradicated by then. It seems most unlikely that it will be.

It is further claimed that virtually the entire committee now backs vaccination, even those who were originally opposed to it, and it was blocked only by Scudamore and the leaders of the NFU. In response to *The Independent*'s request for his current views on vaccination Ben Gill replied that it was 'bilge'. Mr Davies, who dealt with the last European outbreak in Greece in 1996, told the paper that 'had we vaccinated during late March, we would have been spared the loss of thousands of animals and public revulsion at the carcasses across the countryside. However, I had underestimated the almost mystical belief in the superiority of slaughter as a method of control. This has allowed all sorts of false statements to flourish.'

It was also admitted by the government that thousands of people are already eating vaccinated meal brought in from abroad, in spite of the NFU's insistence that people would never eat it. In fact, 18,718 tonnes of beef from vaccinated cattle was imported into this country from Brazil in the year up to May 2001 (DEFRA figures).

During the first week in September, the British Association for the Advancement of Science held its annual conference, this time in Glasgow, at which scientist after scientist rose to call for a change of policy on vaccination. On Thursday, 6 September, Chief Scientist David King told the *Today* programme that he had tried to persuade the government to implement a policy of limited vaccination back in April, but this had been prevented by pressure from 'farmers' leaders'. Some of the impact of this remarkable declaration was, however, negated by his further statement a day later when he told the media he was 'confident that the disease would be snuffed out in the Northumberland blue box zone'. A zone which now extends way outside Northumberland into Cumbria, Lancashire and Yorkshire. He estimated it would 'take a month', adding that 'the models on which such forecasts are made are not precise'. He can say that again. He also insisted that there was no question of introducing vaccination as it would undermine attempts to keep track of the infection by looking for the virus in blood samples. 'If I thought vaccination could be effective in any way, I would be recommending it to the Prime Minister.'

However, in spite of such contradictory remarks, it seems that attitudes are changing by the day, if not the hour. On 8 September, John Humphrys interviewed Junior Minister Elliot Morley on the *Today* programme, and the airwaves were creaking with the sound of frantic backpedalling. Mr Morley has been quoted frequently in earlier chapters, stoutly defending government policy, often in the most tactless fashion. This time, however, we were informed that the government had never *really* been against vaccination, it had always been a possibility as a tool. A number of high-profile people who had been opposed to it were then wheeled out for our inspection. Morley, even going so far as to agree with Humphrys that one of these was former Agriculture Minister Nick Brown.

Next in line were those 'farmers leaders', whereupon Humphrys put into words what vast numbers of people have been thinking for weeks: that the NFU leaders truly represent mainly those big farming interests who stood to gain large sums of money in compensation from the slaughter policy. He also pointed out that the sums lost to the agriculture industry from our not being able to export meat as a result of vaccination (a lynchpin of the NFU case) amounted to £3.75 million, set against a loss to the tourist industry estimated at £5 billion.

On the same day, *The Guardian* published a major feature in which the big food companies, led by Nestlé, were blamed for the pressure they put on the government not to vaccinate. Secret minutes had been leaked to the paper showing that at a meeting at Chequers in mid-April, Blair had agreed a policy of limited vaccination, but the food industry had forced him into an about turn. The paper quotes Patrick Holden of the Soil Association as saying that the industry was on one side of the table, the government on the other. 'Mr Blair said that he wanted to vaccinate but the NFU was telling him that the government could not.' Chief Scientist David King then made the case for limited vaccination. Blair also asked for views from leading supermarket representatives, such as Terry Lee, managing director of Tesco, who said he was in favour of it, after which the rest fell into line as did Lord Haskin.

But scarcely was the door closed on the meeting when intense lobbying began, led by Peter Blackburn, then Chief Executive of Nestlé UK and also President of the Food and Drink Federation (FDF) and Lady Sylvia Jay, a former civil servant at the Department for International Development and wife of Britain's ambassador to Paris. Mr Blackburn's pitch was that a vaccination programme might affect sales of powdered milk. Nestlé is the major producer of powdered milk, with 75 per cent of its ouput going to developing countries. It also exports more than £5 million worth of chocolate, the producation of which is dependent on British milk. A policy of vaccination, said Blackburn, would affect Nestlé's factory in Cumbria. Not to mention the firm's profits. He was immediately supported by the NFU. Faced with this opposition from big business Blair, as usual, rolled over and plans were shelved.

But these revelations, interesting as they are, now look like part of a deliberate attempt to soften up public opinion ahead of a change of policy, the subtext being that just about everyone was to blame for the government's refusal to vaccinate except the government itself. Be warned, you never know who might end up in the frame. Within days of the limited vaccination being dropped, even before the election, the Millbank spinners were at work on Nick Brown and the Cabinet Office was briefing that he would be replaced after the general election. Next in line might well be Ben Gill, a suggestion that was pointed out to him in Professor Vernon Wheelock's open letter of 18 July. His personal stance on the subject, the regular cheap jibes and belligerent comments in response to reasoned arguments, have left him dangerously extended at the end of the twig when the time comes to line up the scapegoats. Indeed it may already have begun. On 9 September, the *Independent on Sunday* 'revealed' that there had been other secret meetings at the end of March and beginning of April when an immunisation plan was drawn up, but this too had to be abandoned at the last minute due to the opposition of the NFU.

So, first it was the siren voices of the food industry, then the NFU that forced the Prime Minister to abandon his plan. But wait, surely it was Mr Blair himself who 'took charge' of the fight against the epidemic at this time?

Speaking at the British Association for the Advancement of Science meeting in Glasgow on 7 September, Professor Fred Brown, who now works with the US Department of Agriculture, whose offers of advice and assistance were continually turned down, warned of an international epidemic of foot-and-mouth. 'While there's a disease round the world, every country is at risk. The US is frightened to death about what might happen if it gets there. I would like to see an international debate on vaccination.' It was still not too late to start vaccinating animals in this country, he continued, starting with cattle and rare breeds.

David Rowlands, a professor of molecular biology at Leeds University, speaking at the same conference, said it was only a matter of time before other countries were affected. 'When vaccination was stopped in Europe in 1991, a lot of people felt this was a disaster waiting to happen.'

Similar views were expressed in the West Country following Professor Brown's widely publicised statements, a round-up of views appearing in the *Western Morning News* also on 8 September. The Countryside Alliance called for academic and scientific expertise to be put at the forefront of efforts to control the epidemic, and in particular the issue of vaccination. Chief Executive Richard Budge said that the decision to vaccinate must 'not be influenced by the vested interests of groups or individuals. The greatest danger to rural people is being caused by the disease having such a long tail. To most people the consideration of restarting export is unimportant. With the winter coming on, the single focus of the government should be getting rid of the disease. We can deal with other issues at a later date.' Ruth Lea, the Institue of Directors' Policy Director, commented that while her organisation were aware of the need to export meat and farm products, its members were mystified by the NFU's approach.

Anthony Gibson, director of the South-west NFU, whose reasoned and sympathetic stance throughout has been a light shining in a dark world, said: 'The vast majority of farmers would go along with it if it would end this outbreak. A lot of the prejudice against vaccination has now been stripped away and it should be looked at on its merits.'

If the government does now opt for vaccination it can blame who it likes for the appalling waste of time, money and animal life over the last few months: MAFF, the farmers, the scientists, big business. But, at the end of the day, the buck stops at the door of Number 10 Downing Street. Small wonder then that there is to be no public inquiry. At the time of writing the number of confirmed cases stands at 2,007 — and rising.

# Sources

EEC Council Directive 85/11/EEC 18 November 1985. EEC Commission 1999: Strategy for Emergency Vaccination Against Foot and Mouth Disease. Report of the Scientific Committee on Animal Health and Welfare.

1969 Report of the Northumberland Committee on the 1967/68 outbreak of Foot and Mouth Disease.
'The 2001 Foot and Mouth Outbreak: Economic Arguments Against an Extended Cull', Professor Peter Midmore, University of Wales, Aberystwyth (2001).

'Foot and Mouth Disease in the United Kingdom: Problems with the Current Control Policy and the Feasibility of Alternatives', Dr Keith Sumption, Centre for Tropical Veterinary Medicine, University of Edinburgh (2001).

'Foot and Mouth Disease: An Evaluation of the Current Control Policy from a Historical Perspective', Abigail Woods, MA, MSc, VetMB, MRCVS. Wellcome Unit, University of Manchester (2001).

'Vaccination Against Viral Disease as Applied to Foot and Mouth

Disease', Dr Ruth Watkins, Clinical Virologist (2001).

'Failure to Vaccinate – A Misjudgement', Professor Vernon Wheelock, former Head of Foot and Mouth Policy Research, University of Bradford. Also his Open Letter to Ben Gill (18.07.01).

USEFUL WEBSITES
www.sheepdrove.com
www.cullmaff.com
www.warmwell.com
www.farmtalking.com
www.epynt-disaster.co.uk

From Sheepdrove it has been possible to download most of the academic papers cited above, along with copies of the Northumberland Report. This site, organised by farmer and ex-publisher Peter Kindersley, was among the first pressing for vaccination. The next three are more radical in their approach but are also very good for up-to-date news, figures, etc, unobtainable elsewhere. The last is specific to what is happening in the Brecon Beacons.

The Heart of Devon Campaign, started by the *Western Morning News*, has now become Heart of Britain.
Website: www.heartofbritain.com. E-mail: heartofbritain@aol.com

Barrister Jeremy Roe, mentioned in the text, has now formed Action for Foot and Mouth, which is proposing to sue the government for loss of business and for damages for negligence and nuisance. At the time of writing instructions are being prepared for Counsel. Website: www.action-footandmouth.co.uk.